DEAD MAN LEADING

V. S. PRITCHETT was born in Ipswich in 1900. At the age of 16 he left school to work in the leather trade in Bermondsey and in the twenties became a shop assistant in Paris, then a journalist in Ireland and Spain, and a reviewer on the *New Statesman* until the seventies. His publications include *The Spanish Temper, The Living Novel, The Myth Makers,* biographies of Balzac and Turgenev, five novels (including *Mr Beluncle* and *Dead Man Leading*), many volumes of short stories, and two volumes of autobiography—*A Cab at the Door* and *Midnight Oil.* He has travelled widely in the United States and South America. He was knighted in 1975 and is President of the Society of Authors. He lives in London.

PAUL THEROUX was born in 1941 in Medford, Massachusetts. He has taught at universities in Italy, Malawi, Uganda, and Singapore, and since 1971 has divided his time between London and Cape Cod. He has published nineteen books including *Saint Jack* (1973), *The Family Arsenal* (1976), *The Consul's File* (1977), *The Old Patagonian Express* (1979), *The Mosquito Coast* (1981), and *The Kingdom by the Sea* (1983).

ALSO AVAILABLE IN
TWENTIETH-CENTURY CLASSICS

Adrian Bell CORDUROY

Adrian Bell SILVER LEY

Hilaire Belloc THE FOUR MEN

Arnold Bennett RICEYMAN STEPS

Hermann Broch THE DEATH OF VIRGIL

John Collier HIS MONKEY WIFE

Cyril Connolly THE ROCK POOL

Walter de la Mare MEMOIRS OF A MIDGET

Peter De Vries MACKEREL PLAZA

Ford Madox Ford THE FIFTH QUEEN

Robert Graves SEVEN DAYS IN NEW CRETE

Patrick Hamilton THE SLAVES OF SOLITUDE

A. P. Herbert THE SECRET BATTLE

Sinclair Lewis ELMER GANTRY

Rose Macaulay THEY WERE DEFEATED

Saki THE UNBEARABLE BASSINGTON

Lytton Strachey ELIZABETH AND ESSEX

Rex Warner THE AERODROME

Denton Welch IN YOUTH IS PLEASURE

H. G. Wells LOVE AND MR LEWISHAM

Edith Wharton ETHAN FROME *and* SUMMER

Leonard Woolf THE VILLAGE IN THE JUNGLE

V. S. PRITCHETT

Dead Man Leading

INTRODUCED BY
PAUL THEROUX

Oxford   New York

OXFORD UNIVERSITY PRESS

1984

Oxford University Press, Walton Street, Oxford OX2 6DP

London New York Toronto
Delhi Bombay Calcutta Madras Karachi
Kuala Lumpur Singapore Hong Kong Tokyo
Nairobi Dar es Salaam Cape Town
Melbourne Auckland
and associated companies in
Beirut Berlin Ibadan Mexico City Nicosia

Oxford is a trade mark of Oxford University Press

Introduction © Paul Theroux 1984
First published 1937 by Chatto & Windus
First issued, with a new Introduction, as an Oxford University Press paperback 1984

British Library Cataloguing in Publication Data

Pritchett, V. S.
Dead man leading.—(Twentieth-century
classics).—(Oxford paperbacks)
I. Title   II. Series
823'.912 [F]   PR6031.R7
ISBN 0-19-281469-9

Library of Congress Cataloging in Publication Data

Pritchett, V. S. (Victor Sawdon), 1900–
Dead man leading.
Originally published: London: Chatto and Windus, 1937.
I. Title.
PR6031.R7D4 1984   823'.912   84-836
ISBN 0-19-281469-9 (pbk.)

Set by Herts Typesetting Services Ltd
Printed in Great Britain by
Richard Clay (The Chaucer Press) Ltd
Bungay, Suffolk

FOR MY WIFE

CONTENTS

Introduction by Paul Theroux   ix

BOOK ONE   1

BOOK TWO   43

BOOK THREE   105

BOOK FOUR   163

EPILOGUE   217

INTRODUCTION

BY PAUL THEROUX

V. S. PRITCHETT, a master of the short story, once wrote, 'the masters of the short story have rarely been good novelists'. But *Dead Man Leading* is an extremely good novel. It has an intensity, a passionate strangeness, that sets it apart from Pritchett's other novels, and except for its language—always the vivid, unexpected word or image— it has little in common with his short stories. In setting and mood it is wholly original. It is certainly the best of the five novels Pritchett has written (though I am an admirer of *Mr Beluncle*), and yet it is practically an unknown book. 'This novel was more imaginative than my earlier ones,' Pritchett wrote in *Midnight Oil*, 'but it came out not long before the war started and that killed it.'

When I read it the first time I was immediately reminded of *Brazilian Adventure*, by Peter Fleming, which appeared in 1933. Pritchett told me that he had been influenced by that travel book—the search for the explorer Fawcett who went missing in Brazil in the 1920s—but that he had also thought of his fellow writers, the restless novelists who, in the thirties, poked around the world inviting privation. He explained in his autobiography that he had read the lives of the African explorers and the accounts of missionaries' experiences ('Missionaries always write down the practical detail') and he found a common characteristic to be an 'almost comical masochism'. It was *Green Hell*, and a biography of Scott of the Antarctic, and—he told me—'memories of childhood fancy' that heartened him to begin writing. He summed up his intention in the novel by saying, 'I attempted a psychology of exploration.'★

★ 'Exploration is the physical expression of the Intellectual Passion', Apsley Cherry-Garrard wrote at the end of *The Worst Journey in the World* (1922). This

It is that and more, and is especially remarkable for having been written before Pritchett went to Brazil. He says he knew the literature of the Amazon and he created his own Amazon: 'I constructed a small model of my bit of the river in the garden of the cottage we had rented in Hampshire.' I am usually doubtful about novels written by people who have never travelled to their setting. 'And he'd never set foot in Africa!' readers say of the author of *Henderson the Rain King*; but no one who has been to Africa has to be told that, and internal evidence in the Tarzan stories show their author to have been similarly handicapped. This is another reason *Dead Man Leading* is exceptional. Apart from the orang-outang with its gulping Lear-like laugh in Chapter 14—the mostly silent *pongo pygmaeus* lives ten thousand miles from the Amazon—the novel is completely convincing. And of course Pritchett himself is a great traveller; he eventually went to Brazil and was able to confirm that he had invented the truth.

The narrative is not hard to summarize: it is the account of an expedition in search of a man who disappeared in the jungle seventeen years before. 'The Johnson mystery', it was called: what really happened? The question is not answered in a conventional or predictable way, because the men on the expedition are of such conflicting character. Indeed, the achievement of the novel is that it portrays an exquisitely physical landscape along with minutely detailed portraits of the inner life of the men lost in it. In one sense it is as concrete as can be, and yet it is also full of subtlety and suggestion.

At the centre of the expedition (but the resonant word 'quest'

---

masterpiece of travel writing, which is an account of the Scott Antarctic expedition of 1910–13, describes vivid instances of severe hardship and a satisfaction in enduring it that amounts almost to pleasure. Another explorer, Fridtjof Nansen wrote, 'Without privation there would be no struggle, and without struggle no life.' This was also Captain Scott's feeling; in one of his last letters ('To My Widow') as he lay freezing to death in his tent on the Ross Ice Shelf, he wrote, 'How much better it has been [struggling against blizzards] than lounging in too great comfort at home.' Pritchett knew these sentiments from Stephen Gwynn's *Captain Scott* (1929).

would be more accurate) is Harry Johnson—it was his father, Alexander Johnson the missionary, who vanished. The Revd Johnson had often been away for long periods. He had four sons. They were proud of their courageous father and dutiful towards their mother. Harry says that his mother brought them up to hate women, and 'It was necessary for each of them to be the missing father—to be immaculate, her husband.' Harry is only in his thirties, but already he is an explorer-hero who has mapped Greenland, the Arctic, the tropics; he is the subject of admiring anecdotes. He has a stubborn and solitary nature. 'I think I may be very different from other people,' he says to Lucy in an early chapter, and soon after she has an equally tantalizing but more specific perception of him: 'The frightening thing was the closed door in his heart and the fanatic behind it.'

Lucy is a fascinating person, and fully human—her decent fear of setting sail in a terrible wind shows just how dangerous Harry's masochism is. The whole of Chapter 3 is brilliantly done, a passionate battle of wills in what is less a love-affair than a struggle for supremacy. Although Lucy does not go on the expedition, her presence is continually felt, and she is often referred to. Harry is as obsessed by her as he is by his father.

What is surprising in the novel is how closely connected the characters are. Even that marvellous reprobate, Calcott, claims a relationship with Harry's father. It is a novel of the jungle, but it is not a novel of savagery—no cannibalism, no blow-guns, no shrunken heads. Indeed, it sometimes seems like a version of English society feverishly disintegrating in the tropics.

The leader of the expedition, Charles Wright, is Lucy's step-father. But Wright was first attracted to Lucy and only after-wards married her widowed mother. So Wright is a potential father-in-law to Harry. Gilbert Phillips, the journalist who goes along on the expedition, is a friend of both Harry and Lucy— and Lucy was his lover before she chose Harry. In the course of the search for the missing explorer these facts, and others, are revealed. It would not be fair to the reader to state everything baldly here, except to say that where relationships exist there is passion and desire; and Pritchett sees the explorer as something of a sexual misfit.

For Wright, the Brazilian jungle has the enticement of a woman, and the tree-tops have for him the look of 'fantastic millinery'. He had seen the Johnson country—'he would re-member the sun upon the wall of trees like the light on a woman's dress'—and regarded it as his own, something virgin and private: 'He had seen its face and its dress. He longed to be in its body. The talk of the missionary's country and the mystery of his dissappearance was talk of a rival and an attempt to enhance her attraction which he could admire as a connoisseur of discovery and adventure but which . . . made him alert for any sign of betrayal.'

The novel is full of rivalries, and it may also be said that nearly every character in this account of an expedition is travelling in a different direction, mingling flight with pursuit, and usually suffering the satisfactions and frustrations that we associate with lovers. This is another reason Calcott is so welcome. He is a necessary element in the Englishness of it all—not a bloody gentleman, as he says. And he is also full-blooded and funny, unreliable, boasting, and populating the Amazonian town with his dusky children. I am inclined to think that it is the inclusion of Calcott that gives this novel its greatness, for otherwise it would be too intense and too sad; and it does not become truly tragic until this odd clown enters the narrative. The seance in which Calcott figures is one of Pritchett's triumphs as a writer.

One reads *Dead Man Leading* and one is struck again and again by how easy it is to imagine this remote landscape. The men and their obsessions seem much odder than their setting. It is one of Pritchett's most visual pieces of fiction, and it is visual in a particular way. It is not merely the clarity and strength of the images; it is, most of all, their familiarity. He speaks of the brown water of the river resembling strong tea, and a sky like a huge blue house; the forest is faint, like 'a distant fence', and the jungle at another point is bedraggled and broken, 'as if a lorry had crashed into it'; there is a creek 'like a sewage ditch' and some birds rising up 'like two bits of charred paper' and a bad rainstorm making 'the intolerable whine of machines' and a forest odour 'like the smell of spirits gone sour on the breath'. It is what English explorers would see—either a kind of tempting purity and allure in the liquefaction of the jungle's clothes, or else a way of discovering memories and establishing among all these relationships that the imagery of the Amazon is the imagery of England. This is emphatic in the last paragraph:

He went with her to the door. Nothing could have been more like the river and the jungle and the sudden squawk of birds to his unaccustomed sight and ear, than the street light-daubed in the rain, the impenetrable forest of lives of people in the houses and the weird hoot and flash of the cars. He went back to his room, wishing for a friend.

It is impossible in a short introduction to do full justice to this novel. It has real excitements—the desertions, the chance meetings, the overwhelming hardships; and it has pleasures—the by-play between Calcott and Silva, the glimpse of the jaguar in Chapter 10, the rainstorm in Chapter 16, the surprising but surely inevitable ending of the quest. And there are the numerous perceptions, almost casually noted, but so telling: 'He spoke abruptly and hastily to Silva, treating him with that laconic contempt which conceals the gratitude one feels to an accomplice' and when Wright lies dying and Johnson watches, 'There enters with the handling of the sick a kind of hatred, a rising of life to repel the assault of evil.' There are many observations like this in the novel. It is an extraordinary book, in all respects precise and wise.

BOOK ONE

CHAPTER ONE

EVERY few hours the captain took the launch in towards the bank and the monotonous voice of the man with the sounding chain quickened. He stood up and shouted the depths. The launch, overloaded with thirty people, men, women and children with their goods, bumped in the brown river water, swung on to the mud and off again. With the threshed foam at her stern she was like a mongrel worrying a dirty mat; then she quietened and the shadow of the forest wall slid over her noiselessly, cool and green like an awning. These stops were made to collect wood for fuel. Since there were thousands of miles of trees, you stopped, landed and took what you liked.

A mulatto came up to Gilbert Phillips and asked:

'Where are you going?'

Phillips did not understand Portuguese. He turned to his companion, the only other Englishman on the launch. He was squatting beside Gilbert Phillips and stared up from under his furry black brows before he replied.

The mulatto gazed over their bodies very slowly. His eyes were like a pair of warm, lazy, curious flies, impersonally moving over the two Englishmen.

Rubber? enquired the mulatto. Coffee? Cotton?

The accent in these parts was different from the accent in his country, on the timber station, a thousand miles to the south, Johnson said. At first he had difficulty. The speech was thick and crumbling.

No, he said, not rubber, nor coffee. Going very far up the river.

The mulatto considered them. The curiosity of his eyes made their white alien skin itch.

Evangelicals? he suggested.

Johnson laughed aloud at this. Then stopped short because he saw, by Phillips' astonishment, that this was his first laugh for several days.

3

'He thinks we're missionaries,' Harry Johnson said.

'That was a near one,' Gilbert Phillips said, by which he meant that it was near because Johnson's father had been a missionary.

The puzzled mulatto smiled. Like many of the other passengers he was barefooted and had sores between his fingers, some in dirty bandages, the wounds of love. He was a frank man with long loose arms and big hands, dressed like the rest in a greasy cotton suit.

Very far up the river (he delicately suggested) there was gold.

So then Johnson recited to him the formula they had fallen into for all questioners. They were going very far up the river, not this river, but another one, beyond the falls, to shoot 'tigers.' It was useless to tell people they were going on a party of exploration. This would sound incomprehensible, become mysterious and therefore suspicious. The two Englishmen wanted to be left alone.

This was the second day on the river. The signs around them never changed. They unwound like a repeated panorama on a reel. There was the screen of continuing and continuing green on the banks, a flow of brown water like strong tea, bark-stained and root-stained by the drainage of the forest, and a sky like the wall of a huge blue house, quite immovable. Sometimes they were in narrow waters and sometimes in wide—this was the only change. At the estuary and in those places where the river widened into an inland sea, the far bank became dimmed to a lean hot blur of forest, faint as a distant fence in a flat country. In the hours when the river was narrow, the water light was suspended in the hollows of the foliage and the launch appeared like a fly fixed in flight between sky and sky-reflecting water. In channels narrower still, where the branches of the trees almost met overhead, one seemed to be passing show cases of fantastic drapery and millinery, in an overheated shop, green unfolding upon green, in the absolute silence. Sometimes there was a break where the cases had been smashed and the trees apparently grabbed out with their rigging of liana trailing after them; and sometimes there were muddy coves where tree ferns made a

4

shade of ospreys. A thatched hut built upon a platform of piles, whitened now the water was sinking in the dry weather, would be in these hollows; and on the platform brown people sat like grubs, half naked, watching the wash of the launch chuckle along the banks of mud. The people made no sign with their hands, nor was there any expression of curiosity in their eyes, nor recognition. They sat fixed and still with the water light wavering over their bodies in their aquarium of soundless shade.

The launch went on. At thatched villages the siren cord was pulled. Men, playing cards, stopped to stare. There was a woman smoking a pipe and all chewed and spat and lay about sweating and talking, their voices rising to anger like fire spurting up and then dying away to a smoulder. Babies screamed in the arms of heavy women. Men stared down blankly at the children. Two young girls with sores on their lips, like dark and sticky honey, had men about them. Their eyes were sad and beautiful, reddened like dark grapes. The sweet-sour, exciting smells of the women, the heavy acrid smells of the men hung under the awning and, with them, the smell of the wood smoke in the funnel, the green fume of the forest and the sun like a weight holding everything down so that no wind blew it all away. At jetties, where the launch stopped every two or three hours, people swarmed down upon her, struggling with their goods, shouting, fighting, kissing and weeping. The quickest to jump were the cripples. Indians stood waist deep in the water. While the crowd struggled on the listing boat, quietly the Indians swam and dived, coming up to watch, then diving and swimming again because their excitement made them dive and swim like this when a boat arrived. Then on again, the forest continuing the wake dragging behind like an old mat, the jetty turning like the hand of a clock marking an hour gone upon the river face, the crowd as small as birds on the shore, but still heard in argument.

Johnson and Phillips moved up to the bows. All day, talking very little, they sat looking at the river and the forest.

'With all these stops for wood we could have beaten this

launch easily,' Johnson said moodily on the second day. 'We could get through the trees to our river'—they always talked of 'our river'—'and save 300 miles.'

'You can't get through the bush,' said Phillips, alarmed by new ideas. 'And anyway what about Charles?'

Charles Wright was their leader. The launch was taking them hourly nearer to their rendezvous with him.

Phillips took off the floppy brimmed hat he had bought at the coast when he was waiting for Johnson to come up from Rio and join him, and ran a hand through his fair hair. He was taller than Johnson, a slighter, wirier, more nervous man with starting grey eyes, a quick biting way of speaking and a mouth a little askew.

On the shady side of the bows the awning had been rolled up. They sat here because it was where their packing-cases had been stowed; here also at night they slung their hammocks. A dozen other people slung hammocks here too and hung like a crowd of bats, but snoring, kicking out their feet and creaking in their sleep. When they were not sleeping or talking they were re-bandaging their sores.

'What about Charles?' repeated Phillips incredulously.

'It's a nuisance,' Johnson said.

This was repeating what he had said at the coast.

A few months before, by the sea, near Charles Wright's house in England, Phillips had heard Harry quietly make the devastating proposal that they should step into Wright's little cutter and sail off the East Coast to Ostend in a full gale. Slow and amiable in thought, Harry Johnson was fanatical and visionary in action.

'We couldn't let Charles down,' Gilbert Phillips said.

Harry did not answer but sat frowning at the river.

They were both in their thirties. Johnson was one of those men whose chests and shoulders seem to be too broad for their coats, whose heads hang forward heavily, the big brown eyes looking upward at some distant place in the sky and the forehead crinkled between the brows like a grave young bullock's. He looked like a man carrying a load on his shoulders, uncomplainingly, so strong, muscular and awkward, soft in voice and

6

thoughtful in every word he spoke, that his gentleness aroused a kind of sympathy. One was touched by the sight of a man so placidly single in thought, so unencumbered, so manifestly self-sufficient, who must always be searching for tasks equal to this strength, who, immoderate in his standards, yet modest to others, could do so much more than other men without troubling very much whether the instinct was competitive or not. The sympathy one felt was for such a capacity for loneliness. He appeared to know so little and to observe so little of other people or to have exhausted them, but without impatience because of that. The Brazilians watched him. Curiously, the girls came near to him, their thick lips parted. He lowered his eyes and turned away, gazing at the river. His eyelids quivered as he frowned against the light. The sweat stood on the coarse skin of his face and neck. They were reddened by the sun.

His head was aching, Johnson said to Phillips: 'Still aching? Like it was at the coast?' 'Yes, like that.' The face twitched, the eyes were lowered. A neuralgic pain in the head, he said. It was the country, Phillips said. There is a period when a new country overwhelms the nerves, nauseates by the myriad new vibrations of its life. 'While I was waiting for you I had to shut myself up in my room at the hotel for two days,' Phillips said.

Johnson turned his head slowly and listened. He had some already indistinct memory of Gilbert Phillips' eagerness to meet him. Gilbert was in an excited state. The food had poisoned him. He had been lying in the hotel for days unable to speak the language, weakened by vomiting and diarrhoea. Now, forgetting what was in his own mind for a moment, Harry remembered with surprise that underneath Gilbert's jokes about his state there was probably fear. When Phillips finished speaking, he turned away again.

'It's this lousy boat', Phillips said.

Then again Johnson said they could get off the boat and go overland. He made this immoderate and preposterous suggestion with his usual modesty.

'But what about Charles Wright?' Phillips repeated.

'Send an Indian to tell him,' Johnson said.

7

Johnson and Phillips got off the launch to stretch their legs when they could. They did not strike off overland. A government boat was at the jetty at one place and a man on the quay had just taken the mail.

'You taking mail?' Johnson asked. Had the English mail arrived?

Johnson tried to suggest to the men on the boat that they should go through the mail and look for his letters. He argued and bullied.

'You'll never get them to do that. We'll have our stuff in two days,' Phillips said.

There had been nothing back at the coast. Four times a day they had been to the post office and then to the consulate.

'It's two months since I left,' Johnson said. He said no more about the mystery. His expression was blank and obstinate as it was always when he was harassed.

Johnson looked blankly and then turned away, saying nothing. He looked over the heads of the people at the settlement. There were big-footed negroes, Indians easily and vacuously laughing, fat men wearing sun-glasses like beetles over the eyes, and pyjama coats. The flies were thick on the refuse scattered over the shore. On the sandy high ground above the flood-level of the water were a few thatched huts and adobe houses. The two men walked across a sunlit stretch. Under the thatch of their open huts, men in pyjamas were asleep in their hammocks, and the women, ragged and scratching themselves, moved to hide in the enclosed compartments. There was just this palm thatch propped up by four poles and the box at one end for the women. In the doorway of one of the adobe houses an Indian lay asleep. Beside him, propped in the sun, a live boa-constrictor was tied neck and tail to a stick. The only sounds in the settlement were men's shouts—from the drinkers in a furious tavern, and then the slow clapping of wings on a roof, like dusty and ragged rugs being shaken. These were the wings of the vultures.

It was noon. Now birds and men and insects were silenced. The greater silence of the country, always felt, crept out like a visible tide, and deepened with the heat. One saw the silence in the trees, things made of rubber and metal; and in the sky the sun flashed faster and faster like a fly-wheel. The striped awning of the launch looked colourless against the blinding river.

They sat in the shade of a tree with their backs to the water. They opened a tin of food; they began eating. It was corned beef, greasy and nearly liquid.

'This'll kill us.' They threw the tin on the ground and flies blackened it.

While they were eating a woman came towards them, walking close to a wall where there were six inches of shadow. She was wearing a straight white dress like a pillow-case from her shoulders to her bare shins. She was a negress or some mixture of negress and Indian. She was tall and lank and young. She came close and grinned, looking with the curiosity of a grotesque bird at the remains of their food. She looked at the bright tin.

'That's what she's after. The tin. Give it to her.' Johnson kicked the tin towards her.

She did not move until Johnson told her to take the tin. Suspiciously she bent down and snatched it. Then she stood up and grinned broadly at them.

She began to speak. She spoke a few high laughing words. Johnson understood a little of what she said.

Why didn't they go to the house where the captain was? she asked.

'We stay here.'

'Good house.'

'Too many flies,' Johnson said.

She grinned. She looked down upon them delighted with discovery. Her eyes were on the remains of the food. In the deep, young, liquid pupils of her eyes, sunken in their hollows, was a dull crimson core. Her arms and her legs stuck out of her garment and her belly swelled under it like a pot. She was pregnant.

'Take it,' Johnson said, pushing the remains of the food away from him. Once more she watched him shyly and suspiciously and then snatched it up. They saw the flies tacked on her forehead and on her neck. When she was safely away she

9

stopped still and stared, smiling with wonder at the two men.

'You go to the captain,' she said again.

'No.'

Her head was like a flattened knob on her bony shoulders, a head drained in the pot of the womb. She put out her hand and begged for money.

'Go away,' Johnson said.

'She won't go away.'

'I'll make her go.'

Her eyes moved quickly from one to the other trying to know what they were saying. She pushed out her belly and laughed at it.

'You come?' she said to Phillips because he had not spoken. She looked sadly at them. She patted her round belly.

'You come?'

The deep single line of his frown was creased in Johnson's forehead.

'Let's get away,' Johnson said after a long time. The only way to rid themselves of the woman was to walk back to the launch at the jetty. They got up and left her.

They saw the negress among the crowd which came down to the launch in the afternoon when it sidled off. She stood there with the same vacuous black grin. Even the same flies seemed to be on her head. There was no change in her at all and she did not look at them.

'There's that bloody woman.'

They both hated her. Johnson, affecting indifference, glanced up at her with a look secretive, intense and almost fearful.

They were glad when the four blasts of the siren went up, impudent and superfluous, and the launch under its smear of wood-smoke took them from the place.

The two Englishmen moved apart now. The heat swamped the lungs. Phillips, the fair one, looked through his dark glasses at the river; the dark one sat with his knees up and his back against the packing-cases under the awning in the bows where the mud-choked water slapped and the iron of the gunwale burned the skin if one touched it. The tar sizzled in the seams.

Johnson sat without glasses, biting his pipe, frowning now that he was alone. If a launch going downstream passed them or Indians went by in their long sailing canoes, thatched in the stern and drifting low in the water like some Swiss Family Robinson on a water-logged raft, his eyes opened with untroubled interest. For a moment he escaped from himself; then the frown came again deeply between his brows, he bit once more upon his empty pipe, his hands twitched and the eyelids trembled over his eyes. The sweat stood like an August rime on his forehead and was in the roots of his hair.

Presently he got up and the crowd made way for him. They looked at him with wonder. The women half-smiled; the men turned round slowly and squared their shoulders and stared after him. He went up to the captain of the launch and spoke to him quietly. Here the heat of the engine was fiercest. Through the noise and the wood-smoke and the sound of rattling screws, the captain said they would put ashore once more for wood. An argument started between Johnson and the captain. Passengers gathered.

Did he want to get off? they asked.

'You should have got off at the last place,' the captain said.

Yes, came the chorus, at the last place if he wanted to get off, that was where he should have got off.

'Now there is nowhere.'

'Nowhere,' many people said. It seemed to give them satisfaction. Shyly Johnson turned and stood staring at the river. Since he stood still without speaking, without listening and without answering, a side show that had shut up shop, the passengers went away. Then he too went back to his place by the boxes and looked straight ahead of him down the broad mud road of river.

'We ought to have got off at the last place,' Johnson said to his companion.

'Off?' Phillips said. 'You don't seriously mean that. We've got to go on to meet Charles.' He gazed at Harry for a long time but read nothing in his face. 'What's wrong, Harry?' he finally asked.

'Nothing,' said Johnson. 'Bloody head.'

Endlessly, fantastic, monotonous, the forest went on lining both sides of a street in silence to see a victim pass to his execution. Johnson's thought ran on: could one break their ranks, get through and escape to the other river, without seeing Wright?

CHAPTER TWO

THEY were still together in the bows. The afternoon was failing and, with the suddenness of all tropical things, clouds had appeared like long indigo tongues out of the south-west. The talk of the passengers had been stamped out by the heat but now began again.

For hours the Englishmen had not spoken. They had dozed and slept. Then Johnson spoke out suddenly and sharply like a man talking in his sleep.

'Take that woman away,' he said. He was lying down full length, with his eyes closed. On his lips was the sullen expression of the sick.

Phillips took off his sun-glasses and looked anxiously at his friend. He looked from lips to eyes trying to read a meaning into their nervous movements.

'What woman?' Phillips said, looking round. Two or three men were standing near but there was no woman.

'The nigger. Take her away,' Johnson said.

Phillips looked about him. The mulatto who had spoken before was sitting near them and said:

'He has the fever?'

Phillips did not understand. The others, near by, turned too.

'He is ill,' they pronounced, staring down.

Phillips understood nothing.

'Are you ill, Harry?' he said. 'God, he *is* ill.'

Phillips was a man who got into a panic at once. He had heard Johnson's complaints during the day. He had heard of the suddenness of sickness in the tropics.

'Is there a doctor?' he said in English. 'Doctor?' Then he made absurd signs, opening his mouth and pointing a finger into it. He looked up at them like an anxious dog ready to fly into a rage.

Afterwards he said, 'Being out of England makes you lose

your hair. Just like the dagoes.' He came back to that word. He was the kind of Englishman who objects to the word 'dago.'

They all shrugged their shoulders and stared, pushing out their lips. It was not remarkable to them that a man should have the fever. It would not surprise them if he died. They said again, 'He is ill.' And one or two new ones came up, pressing casually closer. 'Is he dying?'

'Y-a-t'il un médecin—sur le bateau?' Phillips commanded.

'Médico,' said the mulatto.

'Yes, yes, médico,' said Phillips, standing up. He looked scathing and lordly, his head thrown back and chin up when afraid.

They all watched him with slightly awakened interest but did not move. It was chiefly his fair hair that interested them. Then Phillips thought he was making a fool of himself because now Johnson's eyes were wide open, watching him too.

'How do you feel, Harry? I can't speak the language. Ask if you can for a doctor.'

'Take her——' Johnson said and, with the weary cynicism of the sick, closed his eyes again.

'There is no nigger woman,' Phillips repeated very quietly. 'We left her at the last place. She's not on board.'

He jumped to the conclusion that this was the negress Johnson meant.

'Move back there. He'll be better left alone,' said Phillips, turning to the crowding men. He hated scenes. The group did not move.

Many more passengers came up to see the big man who was ill. They came because he was big and his face was red and because the other one was fairer than anyone they had seen.

They came along to see a man who was bigger than they and dressed in strange clothes, probably in some way rich. He wore expensive boots.

So this is why he wanted to go ashore, Phillips thought. Why didn't he say he was sick?

Phillips invented natural laws to comfort himself: Johnson would be better when the sun went down. He waited. The sun when it went down had the blazing cocotte-eye of some gaudy-feathered bird of paradise over the trees. The prow of the boat cut into water that fell away in crimson and silver swathes. Then the darkness came and the stars. The mosquitoes, thinning out when the launch was in midstream, came in whining clouds round the hurricane lamps when she was near the forest.

He got Johnson early into his hammock. He was shivering and sweating, grumbling at assistance. The passengers thought he was taking their places and protested. The mulatto came and helped Phillips and argued with the crowd. Phillips and the mulatto smiled at each other and talked but neither understood.

'He's all right,' said Phillips. But he was thinking: Suppose he dies? Suppose he dies before we get to Wright?

The mulatto shook his head. Under the awning the air was thick and heavy with body smells and tobacco smoke. The darkness was striped with the dirty yellow light and the confused shadows cast by it and no one could see Johnson clearly in his hammock. The passengers forgot him. When food was being served up and plates clattered and children cried, they were all so absorbed they did not see Phillips roll up the awning to let the air blow through. Through the voices, he could hear the water sheared away from the prow as the launch went on.

Gilbert Phillips was a journalist, thought this country a bit of a sensation in itself. An accursed profession, journalism drives its member to think in headlines, interviews, quotable words and melodrama. Every thought they have is printed and on the front page the moment they think it. When Phillips saw Johnson sick, he saw the headline: 'Missing Missionary's Son Dies in Jungle.' That was for the English press: the little English papers in South America would stress the commercial side: 'Tragic Death of Timber Importer's Nephew.'

But why assume Harry would die? Phillips fell back upon tricking the gods: When the plates stop clattering, Harry will be all right, when the people stop talking, when they go to sleep, Wait until midnight and he'll turn the corner. He'll be better when the first light comes. Quite better when the sun rises.

He longed for the magical rising of the sun and even more for the meeting with Charles Wright, the leader, the custodian of salvation.

'Bloody fool I am,' Phillips suddenly remarked aloud. 'Wind up about nothing.'

So he set laboriously to work to prove to himself that his panic was absurd. He entered into argument and sat dictating telegraphic conclusions and assertions to the black river and the invisible forest. A chap like himself, he pointed out, too impressionable. Always dying. Always waking up and finding not dead. Very humiliating. Great argument my safety this expedition is Johnson. But Johnson sick. That's nothing. He has lived in the tropics for years and is still alive. Cardinal point of my faith: Nothing can happen to Harry Johnson. Why not? Johnson himself says so. Look at his record. Last year he came home on leave to England and no sooner ashore than he is off again. Where? To a nice quiet walking tour in the Pyrenees? Not on your life. Off he goes to Greenland. Says he wants a change from the tropics. The tropics can't kill him. So he goes off to the Arctic to see if he can freeze to death.

Phillips saw his two pictures in his mind: Johnson half naked in the tropics; then, black-out, and Johnson astoundingly appears, on his last leave, in Greenland, a shaggy black head, among emerald icebergs, baggy in Arctic suit behind the sleigh, sitting by paraffin lamps, taking meteorological readings in the snow. The amphibian disabled, he and his companions camp in the closing floes: in a week they haul up the plane before it is crushed. 'There is always something to do,' Johnson said. For fourteen days he and the others tramped up and down in the snow, trampling it down to make a runway for the plane. 'It occurred to me that it would save the sweat of shovelling. It kept us warm and occupied. It is essential to be occupied.'

Phillips knew about it all. He had edited the copy in London as it came through, put in the photograph that no one recognised, made the mistake in the name of the island. 'Sorry,' he said, when the Greenland expedition returned. Found explorers damned sensitive. Didn't understand newspapers.

'God,' they said, 'Thought Gilbert Phillips knew better than that.

Johnson was the man in Greenland and the man in Brazil, the lucky beggar. Passing from Cambridge, by the luck of having a ripe and rich uncle, into the Brazilian timber business. At night, behind mosquito wire, he sat on the verandah of a bungalow, reading *Pickwick* and *Hakluyt's Voyages*; took the slow, happy long view of politics. On leave he was in Switzerland and Austria, Norway and Sweden, sailing, climbing and no politics at all. The world his playground. People mimicked Harry's voice, 'At Cambridge we learned that it is unwise to have women in camp. Things get soft.' And Johnson laughing deeply at himself, but not changing his opinion.

Johnson was action and laughter. He laughed and was laughed at. Sailing, he always fell into the sea. Ropes broke when he pulled them. Losing his dinner-jacket, he appeared at a public dinner in Copenhagen in someone else's. In view of his chest measurement . . . By mistake he eats two breakfasts. In Brazil he wakes up to find a snake asleep on that chest. He is treed by wild pigs. When he goes up-country from the timber station, he is often three months alone.

The laughter of Johnson was deep, soft, shy boom and he would put his hand to his lips at the end of it. He was not a humorous man and it surprised him that people laughed, so he laughed too out of good nature.

Phillips sat stunned with wonder through the night. He could hear the thumping of his heart. The prow of the boat was himself eagerly cutting the flat river water. He had comforted himself. Not he ill, but Johnson! Johnson, the unique and impervious one! Phillips felt draughts of strength coming into him from the night because Johnson was stricken and he pitied Johnson, as a woman would pity a sick lover.

And now, Gilbert Phillips had strength and had quietened his fears. He would go quite calmly over what would happen if Harry Johnson, the erstwhile impervious one, should die. For example, Harry had 'died in my arms.' Phillips wrote letters to England, first to that very difficult woman, Harry's mother. ('Wife of Missing Missionary Hears of Son's Death in Jungle.')

Then he wrote a letter to Lucy Mommbrekke. This was Charles Wright's, their leader's stepdaughter. Harry Johnson had had an affair with her. But how could he write a letter to her? Very quietly and near, as if she were sitting beside him on the launch, Lucy's voice came suddenly to Gilbert:

'Take care of him for me.'

He was back in England in a flash. He saw Charles Wright's house. It was built on fairly high ground among a clump of ever-shivering ash and fixed ilex, with miles of marshland, that were greener than any inland fields, to west and east of it. From all the northward looking windows of the house one saw the marsh ending at a sea-wall half a mile away, where Wright's boat lay in the mud and the land broke up into the reed hummocks of the estuary. Over the ever whistling mud at low tide the birds flew, the redshank and the gull.

Phillips, hearing the water birds of the tropical river remembered it and the week-ends there before he left.

He saw Lucy coming across the marsh with Harry from the sea. She was wearing his mackintosh. Her black, closely curling hair was blowing from her fine, very white forehead in a way that gave her a droll and wanton appearance. Harry and she were lovers. Phillips remembered how she and Harry bickered as they came across the marsh, and smiled at the change in Lucy. All independence and all elusiveness gone, she followed Harry and obeyed him like a slave; hot, outrageous and almost impudent in pursuit of a reluctant lover, it seemed now that she was the trapped one.

Lucy was not very tall, a soft-bodied, lazy and sensual girl. Her voice, which was quick and light and allusive, with all the music of innuendo, gave an animation to the heaviness of her body. The nostrils were broad and humorous, the eyes dark and lively but her skin was the colour of that very pale clay one sees in certain pieces of sculpture. She was one of those women who are not beautiful, but who are illumined when they smile or laugh, like a dark pool when the sun suddenly shines on it. There was something not altogether English in her and yet she was completely English—English from the north-eastern counties too, from where Gilbert Phillips, fair as a Dane, also came.

'When you are out there together,' Lucy said to him, 'look after Harry for me.'

'Why do you pretend you are not fond of him?' Gilbert said.

It was one of the pretences about this love affair which had been so suddenly pursued, that she and Harry did not wish it to be taken seriously. 'But suppose,' Phillips said to her one day, 'you have a child.'

'What a worry-guts you are, Gilbert,' Lucy said impatiently. And in a voice which was milder with an abrupt irony and tenderness she said, as though arguing not with him, but with herself:

'It was ridiculous that he had never spoken to a woman.'

Phillips had almost forgotten now the jealousy he had felt when Harry had become her lover. Phillips was evasive and she was an evasive woman. The wound had been not to his heart, but to his *amour-propre* and a sensitive man is quick to repair *that* damage. And he had a potent consolation to which, if the wound irked on occasion, he could go. He, too, had been Lucy's lover. He carried her beauty upon his body and in his heart.

She stayed with him one night in the early days of Harry Johnson's return from Greenland when Gilbert and she talked of nothing but the man who, so different from themselves, had made this deep impression upon them.

The problem Phillips had to decide in those days was whether he would accept Charles Wright's invitation to join him on his Brazilian expedition.

'I'm like you,' he said to Lucy. 'I'm always undecided.'

'Why don't you go?' she advised him.

'I think I will,' he said, 'if Harry does.'

'Does it depend on him?'

'He does all the things I should like to do,' he said honestly. He told her how he had known Harry when he was a boy.

'When I was ten my family moved to London,' he said, 'and the first thing I noticed in the place where we lived was that a number of boys wore scarlet caps. I always envied them. Some of them used to come to church. One of them was Harry.'

'One Sunday,' Gilbert said, 'I was in church half asleep. The minister said, "Let us pray," and down we all went. All you could see of him was a big head like a dog's above the pulpit. He was like a dog,' said Gilbert enthusiastically. 'When he prayed he was like a dog howling at the moon. And just as I was going off again I heard him say something about helping those in far distant lands, "bringing Thy light to the heathen," and how Alexander Johnson, the missionary, was lost in the jungle—I didn't know it was in Brazil— and would God save him. I rose up on my knees as far as I dared to see if I could spy out Harry in the congregation. I spotted him. I couldn't take my eyes off him. You know his father was never found.'

Phillips' mind drifted and dwelled in the warmth of the past. The truth could be seen and admitted at this remove. 'I didn't marry Lucy,' he thought, 'because I didn't want to, because I liked her too well.'

But when he had seen Lucy arrested by the sight of Harry, when he found they talked of no one else, and then when she spoke, with a sudden, surprising indignation of the chastity of Harry, Gilbert had been aroused.

He saw again his room at the top of the house, the early darkness of London like a soft rubbed charcoal against the window, smelled the cool stale smell of gas which he always noted when he came in from his office. He and Lucy had had tea and now they were standing by the window at the edge of that gulf that suddenly opens in the friendship between men and women.

When it came to making love he had no doubt he was no less reluctant, no less caught in the throat by fear than Harry was, and it was precisely at reluctance that Lucy laughed. He stood beside her. Below the house he could see a wilderness of mews where the wireless played and, rising beyond the mews, were two long black cliffs of flats with the crawling sea of the distant traffic breaking in soft, firm gasps against them. It was the hour of day when people are going home and the lights jump in the windows of the buildings. He felt that he was breasting the whole tide of the city when at last he heard his voice, like some croaking, recorded imitation of his natural voice, asking her to stay with him that night.

Well, would she laugh? Would she be angry? Would she argue with him? All that happened was that she lowered her head and turned to touch the lapel of his coat. 'I had never thought of you,' she had said. Afterwards she said they must have quick, clever minds, giving themselves to nothing, if they were to escape the aching inertia of being in love.

Why did he ask her to stay? Why did he feel sorrow and not desire when she did stay? Why did he see her go without regret and avoid seeing her again, letting the days grow into weeks of silence? Why, too, did she not break the silence? Why did he feel for the first time in his life: That was passionless and wrong? And yet, that now they knew each other and were more deeply bound than ever? And yet again that nothing had happened and they were not bound at all? Why did they never speak about it afterwards, pretending that nothing had happened? And why, when he heard that she had passed from her habit of ridiculing Harry to a passionate love of him, did he (Gilbert) feel such an overwhelming tenderness and joy, the friend who ran to them eagerly to join their hands and gaze with delight at them like a girl?

Phillips sat on the launch trying to seize these things again; but the past is a dim-lit undersea world, leading by impalpable corridors to other rooms lit by the same swaying undersea light, and shapes flick uncertainly before the diver's goggles and wash rootlessly away to greener and even dimmer fathoms. Often he saw no meaning. All he knew tonight was: 'I was Lucy's lover before Harry was.'

Lucy was the link between them, the stream flowing between them, bringing down to him the current of Harry Johnson's physical genius; and taking back to Johnson, though Johnson could not know it and perhaps was indifferent to such a thing, the sober wave of Gilbert's affection. 'He does not know I was her lover. I must watch him for her sake.'

Then Phillips was awakened by the sun. How long he had slept he did not know. Yellow and fresh out of the breeding dews of the forest and the streaked mists of the river, the sun went up over the trees.

Every morning since the coast the sensation was the same. One awoke to feel the forest had encroached a measurable distance nearer in the night. One had slept, but through the night grasses and trees, creepers and bushes had been a breeding fume of green. The sap and all the green smells smelt nearer and were stronger than on any day before, and the rot of trunk upon trunk, deep enough to bury a man, on the forest floor, had reached a richer, more feverishly fertile putrescence. Millions of new stalks had grown, millions of new insects had been born. The vegetation bred like the minutes of time ceaselessly one out of another and every second was overpowering with the know-ledge of the infinite fermentation.

And this was not a thing that one merely watched from the detachment of one's place in the river mirror. The physical encroachment upon one's life, upon one's body and spirit, was a bewildering reality. The skin of the Indians was like a bark and on the negroes it was like the rind of a black fruit. The trees were printed upon the people. And one felt the forest first draw one into its shadow and then set to work upon one's life, and sink one into the jungle of the mind's undergrowth. Something farouche began to burn in the darkness of oneself in the presence of the fantastic glooms and brilliance of the trees.

Phillips stretched his stiff arms and rubbed the red lumps the insects had made on his hands. Shivering and yawning wretch-edly, he looked at the wet decks and the silent hammocks swollen with their fat loads. Only one man of all he could see was awake and was scratching the oily hair of his head and looked with pouched eyes, dazed and waking, at the trees. He blinked, grey-lidded like a fowl. Phillips looked at him. Sudden-ly he imagined himself picking up a gun, aiming, pressing the trigger and tumbling the man among bubbles of blood into the river. The picture was so sharp and real in his mind that he went cold with horror and tightly closed his eyes.

It was strange to him that his first impulse this day, before the eyes of the trees, was to kill, and he was left filled with pride and jubilation by it, after a night of shameful fears and wearying memories.

Johnson was sleeping.

CHAPTER THREE

IMMEDIATELY he came back from Greenland, Harry Johnson went down to see Charles Wright in the country. Harry had heard of the Brazilian plan and was already interested in it; but he was uneasy because for the first time in his life he was being talked of publicly. Reporters were waiting for him at the boat. His photograph had been about. He disliked this. He wanted privacy. It was against everything in his nature to have this public notice and he dreaded meeting people. In Brazil he had got used to being alone.

Lucy Mommbrekke, Charles Wright's stepdaughter, heard he was going to see her family. She had heard about Harry from Gilbert; she had read the papers. She got out her car and drove down to the country frankly anxious to see a 'hero.'

She saw a man dark and not very gainly, like herself, who laughed as she laughed and who seemed to have her kind of incompetence in social relationships. Very conventional and polite on the surface, but unsure, as she was; underneath, a man on his own. He was helpless, too, before her mother's smart guerilla tactics in conversation.

'But, Mr Johnson,' Mrs Wright said, 'you seem to me to be as naïve as my husband. You can't expect to go flying off to Greenland and not have publicity. I would love it. Think of us, poor things, *we* never get into the papers.'

'It spoils everything,' Harry said. 'It spoils the whole object of going away.' There was a smoothing, persuasive, shy tone in his voice, very different from Gilbert's quick crackle.

'What is that?' Lucy asked.

Harry Johnson could not answer.

'To be alone,' said Gilbert. 'They want to be alone.'

'You can't talk,' said Mrs Wright to Gilbert Phillips. 'You're going with Charles too, you know.'

'Is it to be alone?' Lucy asked.

'Partly,' Johnson said. The quietness of his evenly spoken voice, so pleasing with the implications of laziness, modesty and peace, delighted her ear.

'I think,' said Mrs Wright, with her smart, ribald laugh, 'they've got girls.'

Charles Wright smiled. In his boiled shirt he looked like a hairy and bearded goat that had been civilised against its will.

'No,' said Lucy, looking at Harry, 'they've got to get away from us.'

Harry Johnson was shy and not good at this kind of badinage, which had an edge of asperity turned against her husband when Mrs Wright spoke. He was a little shocked that Charles Wright, so restrained and sober, should have married this tart and scribble-voiced widow; but on the whole Harry was tolerant because his real life was not in England. England was a trance of irresponsible amusement. He sat enclosed in his trance, quiet merriness in his absent eyes.

When they all went to bed that night Lucy followed Gilbert into Harry's room. She sat on the bed and they talked for a long time.

'Gilbert's the only one who can manage mother,' said she. 'Wasn't she awful?'

Harry looked at her sympathetically. Here was a woman who really understood his wish to be alone.

When Lucy's father died, and before the second marriage, she and her mother had put on black as athletes put on white, were down on their toes, quivering and bang! suddenly off the mark. They were racing neck and neck. They were racing into freedom from the tyrannical man. Who had won today? Who ran the best? Who got what prize? Then the next morning it was the same. The post came, the telephone rang: bang! they were off again on another day's race. They were all over France and Italy, the pair of them together, and very afraid of each other. Is Lucy laughing at me? Does mother really not know what happened last night? For if Mrs Mommbrekke was the more handsome, the more slender, the smarter, the wittier and more self-possessed, Lucy was, after all, the younger.

Lucy saw the young men appear, watched her mother take them from her. There was an innocence in Mrs Mommbrekke, the innocent cunning of the collector who potters in the antique shops, always picking up, hesitating on the point of desire, and discarding.

Charles Wright was one of the young men. He was not very young, indeed, not young at all. But he was like all the others in that he had come to Lucy first and then, before he had known where he was, he had been picked up by her mother. The race between them was always like this. Lucy decided it, off they went and then, laughing to herself, she dropped out half-way and watched her mother finish the course. But something went wrong with this race. It was in Madeira. Charles Wright was convalescent there. He had just returned from South America. He had woken from his illness with a sudden disgust for travel, nauseated by rivers and jungles and dirty camps. He had been frightened, not in his body or his nerves, but in his soul by his illness, and now he was recovering he was ashamed of his fright and said, 'I'm not as young as I was. I'd better give it a rest.' And so he married Lucy's mother.

There was some satisfaction in knowing that Charles Wright had obviously not known whether he was in love with herself or her mother. It was amusing to be the watchful, ironical spirit in such a situation. But the amusement palled. There was a quarrel.

After this Lucy had to leave. She could not stay in his house by the sea. She must leave them to it.

Lucy had grown up with a love of the difficult and oblique and the appetite grows with eating. She wanted greater and greater difficulties. She would make them out of the simplest materials. She tired of the small difficulties, the little dodges, the trifling intrigues. She tired of freedom. She wished for the heavy chain on the ankle so that she would have to exert heart and soul to drag it along after her.

At first one plays with the chain.

If he married, Harry used to think, he would marry a beautiful woman; if she married, Lucy supposed it would be an apt, accomplished man. They repeated this decision to themselves after their first meetings. She was not beautiful. He was not accomplished.

They spoke about marriage very early. It was Lucy who spoke first. They were at least united in their determination not to marry. His job, he said, made it impossible. He could not take a woman into a timber station in the middle of Brazil—'not a white woman.'

'A white woman!' She repeated his words and laughed inside herself.

'Seriously,' he said, 'it's impossible. You don't realise.'

Ah, here was a difficulty, an enormous difficulty, something impossible!

'No,' she said, 'but what will you do?'

'Nothing,' he said.

The chastity of Harry Johnson was a problem.

'You can't go on like that, can you?' she enquired. 'Or perhaps you can?'

So they discussed that very frankly and came back to the original difficulty, the fascinating, insurmountable difficulty: that he could not get married even if he wished. And this was where they were united, for she did not wish either.

'I have never talked to any woman about these things before,' he said, with what seemed to be gratitude, and she was touched; but he really meant, 'I expect this subject is more important to you in England than we think it in Brazil.'

'You know,' she said, 'I think that is a mistake. You have lived too much shut up in yourself.'

She could have had this kind of conversation with Gilbert, but with the great difference that he would have turned the tables on her; and at the end of it they would have been exactly where they had been in the beginning. But after talking with Harry, neither she nor he were as they had been. Every sentence opened a door—to reveal, perhaps, another door, but that in turn opened.

She went one day with Gilbert to call for Harry at his mother's house in London. In England he always stayed with his mother. The house was a narrow three-storey villa, one of a packed row which had been brisk and pink forty years ago, but

27

which, owing to the badness of the building and the precariousness of English middle-class life after the war, now had a seediness unredeemed by style.

It had some eccentric remnants of respectability. A sun blind was over the door. It hung there on a fixed date in the spring and remained there until a fixed day in the late autumn. Here, after her husband's early death, Mrs Johnson had brought up her four sons.

The sound of Mrs Johnson's typewriter could be heard from the street. She earned some small living now by translating missionary publications. She was a very short, withered, white-haired woman with a high breastbone like a plucked fowl's. She looked with the nervous defiance of a child, as though at first she did not want them to come into the house. Then this gave way to pleasure. Her hair was brushed up as though the wind were blowing up it.

'Come in,' she said in a high voice, childish and emphatic, and she took them into her room. All the time she was speaking she was struggling to put away her look of diligence and fanaticism and to be simple and kind. She strained up to Lucy and Gilbert because they were both taller than she.

Old and heavy curtains hung over the window of the room. The pattern of the brown-and-yellow carpet had been clouded by use. Only in auction rooms would one see a similar agglomeration of worn things. The chairs, the books, the tables, and the dozens of spotted water colours of tropical marsh, Brazilian watersides with alligators and herons in mild confabulation among the reeds, and of tigerish wild flowers of the land where Harry Johnson's father, the missionary, had died seventeen years before—all had the pathos of a family life that has come to an end. Lucy looked at the room curiously, but trying first of all to relate it to Harry, who never talked much about his childhood. She felt the aroused curiosity and sympathy a woman feels when she has come by design into a room a man has closed by the accident of his silence. The Mommbrekkes had been Quakers and she felt a Quaker's sympathy with an environment marked both by a successful struggle against poverty and by the shrewd fortitude and independence of Protestant religion. Puritanism in her own family had been tortured and perverted, but in Mrs Johnson's house she recognised the single-minded independence, the tough if eccentric courage of a spirit which was, in part, her own.

Two of Harry's brothers appeared. He was the shortest of them and their voices were louder than his. They came grinning into the room like a rugger team clumping into a pavilion.

'Hullo-hullo,' they said festively. 'Magnificent! Splendid!' Strong, well-fed, good-looking, they made up to Lucy at once.

'Fine show Harry put up last year,' said the eldest to Lucy.

'Magnificent,' said the other. They glowed.

'Oh, shut up,' said Harry.

And the dead were in the 'show' too:

'That's dad,' said the younger, pointing to a portrait which hung over the fireplace, in the darker end of the room.

Lucy saw the blackened portrait of a man with dark curling hair, but only the face and a hand holding a book were clear. A residue of light was on the cheek and was diffused also over the hazel eyes which looked slightly upwards. The hand was holding the book in such a way that only a raised thumb could be seen.

'Put on the light,' someone said.

But when the light was put on it became impossible to see the picture at all. The missionary disappeared into a mildly glorifying effulgence, symbolical of his own physical disappearance in Brazil.

The young men looked up without interest.

'I met your father. He is an explorer,' Mrs Johnson said to Lucy.

'My stepfather,' said Lucy.

'I told you Charles was Lucy's stepfather,' Harry said.

Mrs Johnson nodded as she adjusted her mind to this fact. Then she said to Lucy, 'My husband was a missionary.'

'Oh, Lucy knows all about that,' Harry said.

All the brothers seemed to want to keep their mother from a tedious subject.

'He travelled in the highest service there is,' said Mrs Johnson, in a kindly voice that quivered nevertheless with the severity of having to make clear to Lucy what her sons would never make clear because they had drifted away from their mother's religion. 'In the service of Christ.'

There was an awkward silence. The brothers signalled to Lucy to excuse the eccentricity of their mother.

Quickly Lucy put down her cup and said, 'My stepfather has often told me about him; he has a great admiration for him. I've always wanted to know what really happened. It would be wonderful if they found out this time.'

Harry, who had been playing with the keys of the typewriter, turned round at this and said quietly:

'We are not going to find out. We are going to a quite different country.'

'Yes,' Gilbert put in. 'Three hundred miles away.'

'It's all been done before. It's finished with,' Harry said.

The effect of this was to give Mrs Johnson's manner an hostility to the two strangers in the family. She waited for silence and then said quietly, 'I have my own opinions. You may think you can get on without God, but you can't. My husband's work was the only work which does not bring evil, hatred and lust. I told this to your stepfather when he came to see me many years ago—before he was married.'

'In my country,' said Harry brusquely, 'missionaries are a pest. Give me some more tea, mother.'

A glum embarrassment settled upon Lucy and even upon Gilbert. The eyes of the three brothers, in their several ways so alive with concern for the next turn of the toss, now had the look of absence. Such discussions, so familiar, meant nothing to them. They were withdrawn into an unconscious contemplation of the pool central in their lives. And yet, whatever superficially they may have been thinking, the fact of the death of their father and its mystery was obscurely present in their minds. They knew, without thinking it, that their mother's frenzied energy and defiance sprang from the fixed panic now seventeen years old, which the mystery of their father's disappearance had caused in her. They knew that it was the cause of their family solidarity, deeper than any general gregariousness, their cult of independence, their difference from other people. They knew it so well that they never spoke of it and were scarcely aware of it. He was an open door in them. They had vision through him. As they had grown up and had become restive in argument with their mother, ripening as she dried up, seeing now a pathos in her narrow energies and opinions, each had privately, unknown to the others, imagined the father—added imagination to what memory there was. Slowly each became not only himself, but the father to himself, in his own fashion. They were themselves and then, added to themselves, some vision seen through the open door. And Harry was the most patient, the most sober and most serene of them, Lucy felt. When she left the house that afternoon it was with the feeling that in him the last door of all was this one of his father.

After this Lucy took possession of him. She pursued in London and in the country. 'I feel lost in England,' he said. 'It seems overcrowded and vulgar to me. That's my fault, no doubt. All my friends are working.'

'I'm a friend,' Lucy said.

He looked at her warm, amused eyes, and was astonished at the truth of her words: she had in some way become his friend. He was her curiosity; utterly impossible to marry—that great attraction. And then he was innocent of women. He was a reasonable man: you could talk to him and he was reasonable enough to agree that it was unreasonable to be innocent of women.

They drove down to Charles Wright's together. In the wiry grass of the sea wall at Charles Wright's house, which looked at them from the cattle-bitten marshes like a mild English face, Lucy led him one day to talk. Wright's boat was the attraction there, so she had to pretend she liked the sea as much as he did. The sound of the sea and the wind was woven into their words in a way which made it easier to speak.

31

Their mother, he said, had brought them up to hate women. It was necessary for each of them to be the missing father—to be immaculate, her husband.

'I think,' he said in his quiet way which was impervious to counter persuasion, 'I think I may be very different from other people.' She laughed.

'But Harry,' she said, 'we all think that about ourselves.'

She took his hand and said, 'Where can you have got this idea that you are so different, Harry dear? You do different and difficult things, that is all. That's why I like you.'

She laughed nervously.

'Enormous hands,' she said. 'And hardly a line in them. How extraordinary!'

She kissed his hand suddenly and then said, 'Look at mine,' putting her white hand beside his. She pointed to a line and laughed.

'That's the man I slept with, I expect,' she said. 'I told you,' she said.

She was thinking of Gilbert, but she had not told his name.

He smiled but did not answer.

'Are you shocked?' she said. 'Do you mind?'

'No,' he said. 'I would be afraid of having a child.' He was looking over the stays of the boat. She was half angry with him for not minding, but half not angry because she knew his whole nature was shocked. This secret narrowness of his, so rare in her generation, a joke among her candid friends whose candour all her own sensitiveness disliked, drew her to him. Her reckless-ness rose to meet this buried, rigid self.

He had brought a new sheet for the boat with him on this day and now he pulled it from his pocket. He made a loop and suddenly he slipped it over her wrist, gave it a twirl over her head and held her arm to her waist. He laughed.

'Harry,' she called out, 'let me go.'

The attack was sudden and had the air of a revenge for what she had told him; but when he saw she was angry, he let her go, and his voice was anxious and apologetic.

'You think your strength is everything,' she said. She was

really very frightened. The power had all been on her side up to now.

He let her go and got up and looked at the shining sea, and said, 'I'm going out in the boat.' He wanted to get away from her because when she was with him he could feel that her life was breaking into his mind.

He could feel her kiss hungry in his hand and her confession burning in his mind.

She was very frightened by his sudden loss of quietness and by the pain coming from him.

The gulls came over their heads as they stood there. In slow wide circles the birds went round, firm and white, and free. He lifted his hand and made comic imitations of aiming at them, like a boy. She saw that pain was nothing to him; load upon load of pain, a tighter and tighter screw of hardship could be forced upon him and he would be impervious—he would even seek it. He would seek the most painful thing. Pain was a transcendent gift. He would think he was handing on to her a little of some ecstatic essence which he had won for himself.

She turned back to the house on this afternoon. She glanced at it to see if anyone there could have seen their struggle.

There was a knot he knew of for the wrists— the Brazilian police had one he had seen and practised. 'Pull you up from the sea-wall by the wrists? Oh! That is nothing to what I could do. Oh! I could punish you for dragging out of me all the things that I have never told anyone.'

The lovely thing about him, Lucy said, was his innocence, his easy way of being satisfied by simple things. He loved Charles Wright's house and the marsh country. He liked walking about in it, talking to the people in the pubs. The lovely thing was his unforced love of everything. The frightening thing was the closed door in his heart and the fanatic behind it. She was really touched by the look of apprehension in his face when he came back to the house with the old sheet in his hands and by the undisguised relief he showed when she smiled.

But the next day he tricked her again. In his heart there seemed to be a desire for sudden and abrupt power over her, hidden under his shy, genial manner. A strong east wind had started in the night and it was blowing half a gale by the morning.

'I'm going out in the boat,' he said. 'Come with me.'

'This is where I let him down,' she thought, 'where I can't live up to him. This is like twisting my wrists.'

They argued about it.

'There is nothing to be afraid of,' he said.

'For me there is,' she said. She was afraid enough to admit her fear to him.

He cajoled and argued. The more preposterous his desire the more reasonable he became.

'There is no pleasure in suffering,' she said. 'I'm not going.' It took courage to say this to him because it seemed to her that if she did not go, she would lose him. He would despise her and that would be the end of it. The end? Well, the end of possessing him. Of course she was not in love with him, that was out of the question. In a month or two he would be gone and 'it is impossible for a white woman to live in the jungle of Brazil,' impossible for a civilised woman who has lived all her life in cities. So there was no question of love.

She said she would come down after him to see him start. The boat was floating now, straining at its moorings under the clapping sail. In the sound of the wind and the crackle of the sea, they had to shout to each other. He took off his shoes, rolled up his trousers and waded to the boat. She stood on the shore, forgotten, while he crouched in the boat, going round the knots, crawling into the locker in the bows for the foresail. She sat on the grass wall and watched him. At last he saw her.

'Come and have a look,' he said.

'I can't.'

But after a while she took off her shoes and stockings and, lifting her skirt, went out silently. 'Ah,' he said politely. 'Wait there. I'll help you.' He got her into the boat.

'This is a bad wind for this coast,' she said.

She sat still until her teeth began to chatter.

'I'm going back now,' she said.

He didn't hear her. He was heaving the mainsail to its full height and as he pulled he was looking at the channel. Then the foresail was up and the canvas snapped like a machine gun, deafening them. He was in the bows. He was at the stern. She was looking at his crouching and bending body, smiling at his hair blowing on end. This made his ears stick out in a comical way. Her eyes were dazzled by the light on the superb white flank of sail. He ran towards her barefooted—to the stern again and she was amazed at the excitement in his eyes which, as he took the tiller, half closed to a sudden quietened serenity.

'Harry,' she exclaimed, 'I'm going now. Help me over.' She sat on the gunwale.

He did not answer, but, like a laugh from the sea, there was a long, galloping chuckle of the water against the bows, a hard slap and a sound like the tearing of silk as the cutter heeled and the boom slipped over to the full spreading power of the wind. The boat was sailing. He had tricked her.

She said with quiet anger at him:

'Put me back.' She pulled his arm at the tiller. The boat swung and then lay over once more with a sickening swerve.

She was furiously angry.

'Harry, put me back at once,' she said. 'I told you I'm not coming.'

'I'm awfully sorry,' he said. 'I thought you meant to come.'

'It isn't funny,' she said. 'You're being a bore.'

Soon they were clear of the channel and were sea-ripping where the deeper water opened. They went on the farther side of the estuary, and here the wind came in larger and more sudden gusts making the water pour along almost level with the gunwale. In the east the sky was very pale, almost white as if there were a dust-storm in it. The sea was looser and larger and the boat rushed along with a slow rise and fall like the back of a galloping horse. The waves ran back. They were running over the mounting neck and dipping heads of this herd of waves.

'Put on my coat,' he said, pointing to the one she had brought for him.

There was nothing there but the loose, brilliant mass of the sea, pitted like the craters of a molten planet. He looked ahead and did not answer her questions. He was glad the sea was getting rougher. He had no plan in his head but to get out into the rougher water.

He was like the stone; the wind tore at him but nothing of himself was yielded. He was granite. He crouched, fixed in the hard gladness of another world where the white foam rose like weals on a flogged back that heaved and sent out howling groans. He was in a world of sensations and tortures and difficulties, a spiritual world of obstacles sought and surmounted.

The northern arm of the land had now dropped away. Under the punches of the wind it was impossible to reckon how much time had passed. All that had been broken up. There were no other boats in the sea.

He was thinking how pleasant it would be to go on across the North Sea on a violent day, dazed and deafened and with every nerve tight like the sheet, in the brutal cleansing sea and to find himself in Holland. He would drink a lot in the evening.

If the boat had not been there Harry would not have come down to stay. The sail swelled like a breast in the unreasonable wind, the hull swerved and jumped like a living thing perilously difficult to manage. It had its sudden exertions when it seemed to run away out of your hands and then its times of docility and the caprice that makes you smile with pleasure. Oh! she prayed. If Charles had only sold it. It was too madly engrossing. Yes, she said, it is mad, quite mad. Look at it now; he can't manage it. It flings about just as it likes.

The waves grew larger against the clear sky, the skyline capsized. The bowsprit rose up. The cutter shook under the weight of water that came down like gravel out of a tip, dug and shovelled at it. Yes, it was laboriously mad.

An hour passed swiftly. A second and a third hour went by. She pleaded with him.

To have a moment's silence and stillness! Her breasts were cold, the nipples hard with cold, her eyes streaming, heart gulping in her chest. She turned her back to him so that he could not see her teeth biting her lips and the sick shame of her fear and the struggle of her nerves to surmount it. And she prayed. She turned her back so that he would not see her pray. But if their eyes met she smiled to please him.

'I could not let myself love an ox or a mule or a gorilla,' she reflected; and all the time was busy with thoughts about him, winding, twisting, tying and knotting them, like a woman trying to tie a huge unwieldy parcel with lengths of fine string which break at every moment. Her feeling for him was like this; unwieldy and unmanageable. She fought to assert that there was no blank, helpless chaos of feeling in her heart.

They returned. Four hours must have passed.

At last the squalls of the height of the tide became fewer as they sailed back and after the open sea, the estuary was quieter. 'Now,' he called. She caught the mooring and hauled it in and the sail rapped out like shots as the boat turned with the wind.

There was no sign of anyone in the garden of the house or the marsh, only smoke rising from one of the chimneys. She stood on the swinging bows holding the pole loosely and suddenly it slipped from her hands. 'Oh, isn't that like me!' she cried out.

The pole floated slowly away.

The next thing she saw was Harry pulling his shirt over his head and stepping out of his trousers. His body, so nearly white with the sallowness of pale sand, was bowed in the wind, but in a second or two it straightened and he dived into the sea. He threshed his way after the pole and more slowly came sprouting back with it and the strokes of his water-shortened limbs were blunt and vigorous.

'You go ashore,' he called. For he had realised only then that she was there and he was comically naked.

You look like a rat,' she laughed as she jumped to the shore. When she turned he was out in deep water again and then clambering in. She felt a wave of protective craving for him, as if he were a child and she was his mother, when the white unsheltered body clambered up the side of the boat. She watched the hands grip, the shoulders swell and then the long bow curve of his back to the small cold-coarsened buttocks which made

37

her eyes dance with amusement. She felt a sudden need to rush out to him and to kiss his back, a rush of feeling disordered and heedless. He levered himself over the side and sprawled into the boat and hid from her sight as he groped for his shirt. But she was buffeted and defeated in mind, overcome and bullied in will and in a temper with him, still more with herself. She had been frightened for herself and for him; she could still feel the fear for him tearing in her.

When she got to the house she went upstairs to her room. After the uproar of the sea she could feel all the eventless afternoons of her life standing in the passages of the house and she seemed to be walking through day after day. As she unhooked her skirt in her room she looked out of the window to see if Harry was coming. He was walking over the marsh with his slight crouching, plodding gait as though he were going on for miles and miles. Often he looked back at the sea. She took off her skirt and her blouse and lay down on the bed.

He knew, she thought, that I was afraid. Not to fear is everything to him. After a time—perhaps she had fallen asleep— she heard his voice. He knocked at the door and looked in.

'Oh, I'm sorry,' he said in his absurdly polite voice, seeing her half undressed. He shut the door and went. She sat up and called him back. He came back to the doorway, glancing back in case anyone should catch him talking to her in her room. He was looking away from her to the window and the sea.

He stood red-faced, open-mouthed with a hungry pleasure in the sight of the slaty water. Then he went away. It did not occur to him for a moment that she had suffered. The thought came to him, after he had closed her door, that women were continually ill. He passed the rest of the day uneasily, restlessly going to his chair and sitting there inert and bored. The afternoons in England when there is nothing to do are boring. He wanted his holiday to be over and to be back at his work. There were times in all his English leaves when he wished to be out of the country and at work. This leave had been so unlike the others. So many friends had gone away or were married and had children. With their freedom lost, the men's will went too, it seemed to him. As the years went by he had fewer roots in the country, it would become worse as further years went by. Late in the afternoon, mystified by his restlessness, he went down to the sea-wall and stood staring blankly at the water. It was grey and inexpressibly dreary, thrown down like an old too much trodden rug. Too many people had seen this water. He felt a deep longing to be away with Charles on the river so few had ever seen. He reckoned up the days to his departure.

That evening, unknown to himself, he was good company. He talked at dinner about his adventures. Even Mrs Wright listened without mockery and Charles Wright checked his own talkativeness. They were delighted with him. The reason for his gaiety was simple: already in imagination he was among the things he liked. He was not in England. 'Well,' said Charles Wright as he said good-night to him in his room, giving him a quizzical smile, 'one gets soft here. Women in camp!' He looked through his glasses at Harry, like some seigneurial goat.

Harry undressed quickly and got into bed. His clothes were thrown anywhere, on the chairs and the floor.

He was used to being happy but on this evening he was happier than he could ever remember. He was one of those men who swing from intense action into a warm and dormant content, who regard their jobs with not too much ambition. He was not complacent; he knew he was lucky. It was his luck which made him happy; he believed in it; he was as intimate with his thoughts as with a mistress. Rich or poor, he would have this fine golden wire of luck in his life, the one string that would not snap. He thought it 'sheer luck' that he knew Charles Wright, luck again to be going with him. It was luck that he was different and alone.

He was in the midst of these thoughts when his door opened and Lucy stood there. 'Hullo,' he said.

She moved a little way into the room and only then he saw she was in her nightdress. The blood rushed to his temples and he sat up fixed and cold with wonder at the sudden sight of her body. In two steps she had broken the bemused amiability of his life and stood in the midst of his dreams and desires; for he could not believe at first that she was not one of the women he had dreamed of, mysteriously come to his waking eyes. Not for a moment did he think of her as the woman to whom he had been speaking ten minutes before in the room below.

'I have hardly seen you this evening,' she said.

She came barefooted across the room towards him. There was a boldness in the line of her body. It was solid and warm and the breasts caught in the gown were full and lifted. He could see the dark nipples. He lowered his eyes and when she spoke to him he was without breath to answer. She was not smiling; there was even a sullenness about the dark brows and the lamplight gave a hollowness to her eyes and small shadows to her lips. The candour of her body had something in it that was lovely but terrifying.

'May I come and see you?' she said.

She spoke in a quieter and more musical voice than she usually had, but the body came between the words, solid, firm and peremptory, and underlining everything she said with resignation and sadness.

She sat down on the bed and first of all she laughed at the untidiness of the room.

'I've just come to see you,' she said.

'Tell me,' she said. 'Are you looking forward to going away?'

He was unable to stop looking at her.

'In a way,' he said, 'there is no choice.'

'I shall be sorry when you're all gone,' she said very easily.

'Charles and Gilbert will be back within the year,' he replied. He was looking at her white round breasts which fell forward as she leaned towards him. Sometimes her own eyes were unsteady like his.

'Yes, that is true,' she said.

'Gilbert is a good sort.'

She looked at him closely. Was he trying to find out about Gilbert and herself? Did he know that Gilbert had only a month before been her lover?

'Amusing,' she said.

'I thought,' he said awkwardly, 'you and Gilbert were engaged or something first of all...'

'Good heavens, no,' she said. 'We are just friends.'

'Decent chap,' he said.

She leaned over to him and suddenly put both hands into his hair and pulled it.

'Here! Lucy!' he said, glancing anxiously at the door.

'What is it?' she said, sitting back and looking there.

'Charles ...' he said.

'What about him?'

'He might come in.'

'Charles,' she said gravely, 'has gone to bed.'

'Why,' she said, with a little bitterness like her mother's, 'are you worrying about Charles?'

He was joking, but he was perturbed at the thought of Charles finding her in his room. He hated her really for being there. He was terrified of her nearness to him and when he spoke his voice was choking and small.

She got up and to his relief went to the door, but she had gone simply to close it. When she had done this she stood hesitating in the middle of the room. She was trembling and shy under the turbulence of her desire, which now she could not hide from herself by any subterfuge. It was there like arms drawing at her waist and her legs, padding like an animal in her heart, music swimming in her breasts. She could feel in her hands the brush of his hair on her hands. And because she had exhausted even subterfuge she felt naked and exposed and reckless. Her nostrils dilated and there was a burning weight on her back. She began to pick up his clothes from the floor and put them on a chair. Vehemently she shook out his trousers and folded them, hung up his coat. Then she faced him and marched towards him and stood over him. He could feel the warmth coming from her. She put out her bare arm and took his hand and held it tightly. She could see the startled look on his face, hear the breath coming from his parted lips as she leaned over him until her breast touched his lips and his face. She kissed his forehead. Then, suddenly helplessly weary, she sat down on the bed beside him and ran her fingers into his hair.

'You are a lovely person,' she said very quietly. 'I shall be sad when you're gone and it will be how many years before I see you again?'

He looked up at her.

'I don't know what to say,' he said in a lost voice.

'Don't say anything. Don't worry about not loving me.'

There—she had said the word 'love.' He gazed at her speechlessly.

'Oh,' she said, 'it is cold.'

She was indeed shivering with the coldness of the room. He could feel the coldness of her bare arm when he put out his hand to it. Her arm was white and round and soft and this softness seemed to flow through his hand and into his body and to make her beautiful. She was beautiful now, and when she had come into the room she had seemed to him terrifying and ugly. It was the ugliness of his revolt and panic. Beautiful or repulsive? Ecstasy or horror? His mind was lost.

'Warm me,' she said 'for a little minute. I must go, but warm me. Oh, you're cold too.'

He felt like a man drowning amid the last choking, dying images of his life and Charles Wright was the recurrent one, flashing by, portrait after portrait of Charles Wright, the leader, and then this last one, thrown in suddenly and inexplicably: of Charles Wright standing with Mrs Johnson before the painting of his father, in their house.

BOOK TWO

CHAPTER FOUR

LOYALTY, adherence to a code which respects the privacy of another's soul, and then a dislike of causing pain and of starting quarrels and controversies which may go on for a lifetime—all these virtuous motives work in only too well with a motive less exalted, like the fear of libel, to prevent travellers from writing true histories of their expeditions. Those who have read the two or three accounts which have been done of the Wright expedition must have been struck by their extraordinary lack of background. The speculations and self-important conjectures of the Englishman, Calcott, from whose house the party set out; the irresponsible fantasies of Silva, the Portuguese, who was also there—all have added a specious kind of mystery to what was already sensational enough in itself, without really adding much enlightenment. And those who examined the expedition from a professional point of view, criticising Wright here, taking sides for and against Johnson there, or insinuating that Phillips knew all the time more than he would tell, are not very much more satisfactory. Phillips, it is true, is the source of the most reliable evidence and the book that was written from his diaries is the best; but the diaries of Phillips are the work of a perfect egoist. He pours himself into everything until it is three parts Phillips and only one part itself. He is concerned only with his own sensations. He was one of those sharp-tongued, clever men, whose humour has an intimidating rasp to it; no one seems more sure of himself; but as a diarist he reveals himself behind these defences as full of panicky introspection and self-pity. He seems to have cultivated an appearance of debonair instability in order to hide the dreary, maudlin flats of commonplace that the monotony of continuous fear had made in his soul. A young man should not, perhaps, be judged by his diaries; Phillips, all the time he was in Brazil, was consumed by worry about money. He had thrown up his job, he had put his savings into

45

the journey, he did not know how he would live when he returned. This was his real act of courage. The remarkable thing is that, in this fear, he seems to have been sustained as much as cast down, by being in the company of secure men like Wright and Johnson. Through them he was able to become courageous.

But Phillips—and his diaries show this—was quite unaware that the Johnson he met at the coast and with whom he travelled on the launch to meet Wright, was not the Johnson he had known in England. He records merely: 'Johnson sick. Surprised it is not me.' He is rather pleased with himself. He has no idea that he is travelling with an obsessed man, that the man who is sitting beside him is not the reasonable, sane creature he knows, but one who is drowning in a chaos and clinging like a fanatic to one ridiculous straw.

Yet Phillips was not without sensibility and what he chose to put in his diary and what he chose to leave out or did not think important, is his affair. He knew when he met Harry Johnson at the coast that Harry was depressed about Lucy and particularly because she had not written to him. The affair had been broken off, for as Lucy said, 'We went into it with our eyes open. We know it can't go on for he can't stay here and I can't go with him,' and Lucy's common sense was in action once more under the guise of nonchalance and elusiveness. She was making the break complete by not writing. She was leaving Harry without a word, as she had left Gilbert too. Lovers have a greater dramatic sense than ordinary people, which frequently leads them to over-do their gestures. Lucy was parting with dramatic thoroughness. Gilbert would have been pleased to talk about Lucy with Harry and to have attempted consolation; if he had done so many things might have happened differently; perhaps the whole history of the Wright expedition might have been changed. But no one knows the critical actions of his life. Gilbert was in the position of being Lucy's confidant and not Harry's and he did not speak.

An obsession, fantastic in origin and rooted with daily increasing intricacy, had grown in Harry Johnson which a few words from Gilbert or a letter from Lucy might have dissolved.

Letters were the desire on the one hand; not to see Wright, to avoid him at all costs, was the other desire.

'His lordship's asking for his mail again,' said Calcott, the Englishman, when the river launch arrived at the town where Wright was waiting and they were at the house.

It was sundown. There was one of those crimson, yellow and purple sunsets like the paint of a theatrical backcloth over the scrub and not delicately done but daubed and running. Some red spotlight had caught the few adobe houses of the town, all colour and sound were exaggerated.

In the brilliant heat of the morning, Wright had been waiting with the crowd on the wooden jetty, standing erect and trim with his grey beard in the air, his arms folded and his white hat over his eyes, like a small, shrewd king. Among the raucous, struggling, glaring people, he alone looked subtle, some skilfully made chessman with a few definite, quiet and powerful moves. A lank and wasted Englishman of fifty, with an uneasy mouth, already misted with drink and as bald as a vulture under his hat, stood uneasily beside him. This was Calcott. They spoke quietly, under the shouting and the squealing of animals, to Fhillips down in the launch.

'Harry's sick.'

At last Wright came aboard. Johnson was standing by the cases and Wright walked up to him. Johnson's thick slow smile was on his face and his lips were parted. He put out his heavy arm. It was as limp as a bell-rope when Wright shook hands with him, and Johnson, making a half-angry effort to pull himself together, had his dark eyes in a fixed drugged stare. A veil of fever hid Wright from him. Not to see Wright!—the fever half achieved that for him.

'I'm all right,' he said.

They sat on the cases in the sickening heat. Johnson was most agitated about the canvas boat that was packed with their cases. Sweat ran down their necks and messed the hair on their hands which dampened their knees. Wright chatted in his easy, gay, unperturbed way about his journey and his preparations, but watching Johnson closely all the time. Johnson's strained eyes lowered and his eyelids trembled. His hands closed and opened on his thighs. He listened in a trance, opening his mouth as if he were going to speak but saying nothing. He was impatient to go ashore. His neck was bent as if he had a load on his shoulders and he looked like a rebellious, helpless boy who is being pushed along by a father whose temper is up.

Calcott's house was among others a hundred yards away on a sandy bluff, and this hill finished Johnson. He insisted on walking, but Calcott one side and Wright the other had his arms. The moment he got to the doorway he fell heavily forward. A dozen young chickens ran squeaking into the house before him as he fell. Wright shot out an arm and saved him. Wright was proud of his small hands and his slim arms because they concealed a wiry strength which could catch and hold a heavy man in his fall. Johnson vomited.

They got him into Wright's room and put him in Wright's hammock. Women and children came out of an inner courtyard to cry out that the man was sick. Someone closed the shutters and the room was dark. He lay in the hammock under the heat of the day, which pressed like the chest of a hot man against the shutters, not knowing how long he had been there. He lay dumb and stunned as if he were being borne sick on a litter on one of those long forest journeys in which hour after hour one is travelling down green, skyless tunnels. There was sky now because he was better, a little sky. Only one fact seemed to him important: he had met Wright and nothing had happened. There had been no disaster. What was he to make of that? That Wright did not know? That he knew, but was silent? There were still no letters. This was the most important point. The news would come in the next letters. Then Wright and everyone would know everything. He would see the genial smile passing from Wright's lips, the friendship passing from his eyes.

He had been driven to working it all out again, to go over the details and see where he had gone wrong. Not once but a hundred times, on the river. He had been two months in Rio. A month gave time for letters from England. No letters in Rio. Well, back at the coast there was even more time. To the post office and then to the consulate, the long wait in the afternoon rains: back over the steaming pavements where the flower-petals, torn down every day by the afternoon storm, were a smeared confetti of mock marriage, into the cool courtyard of the consulate. Shy of asking again and again at the consulate. Sending Phillips in. No way of concealing from them. The consul would know in time. The little half-caste at the post office with the cigarette-stump stuck to his lower lips who sat on his stool scratching his leg—he must know. Police censors opened letters, were amused by private news. Phillips would know.

And before that Mrs Wright, scandal-mouthed among her pictures in England, would know and the newspaper reporters who took his photograph at Southampton would know, and by this time Wright would know.

The waves of fear began to break over him again, running up his limbs and breaking with an awful soundlessness in his head. One by one they came and after they had gone he felt dull pain in his waist and his bowels contracted. It is the silence of the mind's convulsions that is unbearable. He lay not one man but two or three, like wrestlers in conflict, too confusing to distinguish in the tangle of their straining bodies.

He longed to talk to someone, but if Calcott or Phillips or Wright himself appeared, his expression became obstinate and resisting. He showed them he was able to manage his affairs alone. When his friends had gone, he wished he had spoken. Yet letters, this idea he had got into his head, were fundamentally irrelevances. Under it all was a fear cutting far deeper than anything else:

'I am losing my nerve. I was afraid of the water, afraid of the sun, afraid of the trees.'

And Wright's words came back to him:

'Women make you lose your nerve.'

So if Wright said to him, as he would be bound to say when the letters came, speaking like a father, 'You've lost your nerve, Harry,' he would have to reply, 'Yes, it's gone.'

'I know. You've slept with a woman,' Wright would answer.

49

'I had a letter. There's a scandal about it. You slept with my stepdaughter. She is going to have a child.'

'Then it is true what I feared,' he would have to answer. 'She did not write because she was going to have a child.'

'It is in the papers,' Wright would reply. 'That's ruined this show.'

That was the obsession: You have been Lucy's lover, she is going to have a child. She has not written because she is going to have a child. She is hiding it, telling only a few people because of the publicity. At Southampton when he kissed Lucy goodbye, some camera man snapped them, 'Missing missionary's son says goodbye to friend, probably to fiancée.'

Still—one is not as innocent as all that. It is upsetting if you get a girl of your own class in the family way, you feel a fool. Good heavens, people say, how very careless! Couldn't they have been more careful or couldn't they have waited? One does not say, 'I have ruined a woman,' like some raffish prodigal in a novelette. Or does one? Is the heart a little Bethel—some sectarian organ which has lost its religion, the texts on the wall and the Israelitish fantasies, but not the fear and the guilt which were the root of the religion? Still, why invent a child? What is the reason for that?

He remembered the blast of the ship's siren and the echoes going down the iron roofs of the dock sheds, as a signal of phenomenal freedom. He waved to Lucy on the quay. She was pale and stolid with wretchedness. There was no smile on her face. Her eyes were shadowed. Rain was falling. He felt an involuntary tenderness for her now that he was going; and with it a feeling of elation and horror. On his last night in England he had expected peace and time to collect himself. At his mother's house he felt guilty because he had neglected his mother on this leave. He had wished to be alone with her. But Lucy, impetuous and desperate in her passion, had insisted on staying at the house.

'Lucy!' he said. He hated her for coming into his room on this last night. And in his mother's house. Once more this hot muddle of love, her hands on him, her breasts against his mouth. Love forced out of him. She stood at the top of the stairs calling to him, careless of who heard, dark-headed, in the yellowish light. The negress on the river had brought this back to him.

At the time he had no clear judgement. He remembered the delight of desire and a glutted repugnance. He had never dreamed of a woman like Lucy. The women he dreamed of, for example, had fair long hair and walked barefooted in a cool landscape of tall flowers. Or in palaces. And he did not know when he had first feared that Lucy would have a child, it was a fear you do not name—the fear had perhaps always been there—or how it had become a certainty for him. It was suddenly a certainty. After the long, serene *ennui* of the Atlantic voyage, when one lived ineffably suspended between sky and sea for free day after free day, it was suddenly a certainty. The sight of the first land, the bay at Rio, might have brought it. This was the first experience of her haunting quality that he had had, for on the ship he had not thought about her at all. She had gone entirely from his mind. But when the ship approached the bay at Rio and the engines quietened as the chaotic coast opened with the brilliant order of a peacock's tail, its momentous and gorged profusion, a weight settled in his heart.

He stood at the rail with the other passengers who, like himself, felt the power of the sliding coast and the beauty of their quiet entry, and he felt the quivering satisfaction of familiarity and recognition. 'A year ago I was here.'

This sudden mark on the calendar arrested him. The timeless delusion of the sea had gone and he knew that he had changed. As he waited in the customs house he thought that here was land where Lucy was not. This gave him the shock of passionate craving for her, tumultuous and overwhelming. He was burdened with love. At the hotel he sat in misery, listening to the electric fans, the resident voices came back to the wearily familiar smells. He walked miles through the hard streets looking at the bay and the lights on the water like cheap jewels on a dark harlot. A storm broke—he had forgotten the storms—and lightning carved up the heat of the night like violet knives. How was he going to endure not being with her?

The next day he went glumly to the office of the company. The thought occurred to him then—and it had recurred on the river; it was one of those master thoughts that strike in the soul, hour after hour, like a clock striking—how was he going to live alone up-country again? He was afraid. It was the first of those waves of cold flame that came over him and remained licking as if the cold fire had caught on to something in him. How was he going to live alone? He had lost his capacity for solitude.

Lying sick in his hammock, he said again, 'I cannot go back. I have lost my nerve.' As Wright had said, 'Women take a man's nerve.'

Seated at their desks made of the beautifully grained mahogany of the forests, were all the men he knew at the office.

'Johnson's back.'

The pens dropped and the typewriters ceased. The Englishmen came effusively; the smart, delicately mannered Brazilians sat staring, waiting their turn after the English effusions. The smell of the sawmill and the forest was civilised here. He shook hands. They were decent fellows. The handshakes of the Brazilians were longer. Laughter, because an office-boy running upstairs to the manager's office, pelted blindly into him and sent a sheaf of letters blowing down through the banisters. A new Johnson joke started: 'He's hard. He broke the office-boy's neck.'

But by the end of the day, there was this:

'It is not like the last leave. The place gets on my nerves.'

It was one of those many grey days when the air is warm, passive and electric. The sky was like cement. The white houses, terraced in the city and the scenic railway suburbs, stand out like hard set faces and their open windows are like square perfectly still eyes. Where had he seen that expression of blank satiety before? All those houses—and this was startling—were thousands of Lucys, turned to sad sallow stone on the quay, looking over a dirty and widening grave of dock water to a departing ship. Then came the first turn of the screw of revulsion: he was linked to Lucy. She could be a house, a hill, a tree. She was not a woman, she was an idea. It was marvellous that she could come like that over the sea and stand before him; marvellous in an evil way, that he was chained to her.

'No,' he said, 'it is over. In five years a lot can happen. I may marry. I might marry a girl like the one who was in the boat coming over a year ago. The red-haired girl. Girl and woman are different things: girl is flower-stem like the field of yellow irises I dreamed of. Woman is something else: too full like Lucy. She will marry. We discussed it all. We said we would not marry. We accepted that. We are free.'

It was then as he walked past the massive negro porter standing by the awning of the hotel door that a voice planted a seed in his ear:

'Suppose she had a child? Then you are not free!'

This time last year ... he was by the lake where the flamingoes and the duck went up in a cloud as thick and noisy as the leaves of a towering pink tree with the wind in it, and the reflection of the immense flight raced under them in the water like a snake. The smoke of his gun blew away in the clear air. In the evenings at that season the sky was clear and greenish. This was in the higher forests among the flat-topped mountains. He rode down in the cool, dew-dazzled mornings to the timber station by the river. Black smoke came out of the tall iron chimney and he heard the saw in the mill, singing, screaming like the sound of a torn sash and dropping abruptly to a dull hum before it began again on the tree. There was the smell of green timber and the tang of resin. The boards rang down as they were piled. In the evenings he went back late to his house for there was nothing to do but work and sleep. He played the gramophone. He read. The telephone bell rang. He would look up from his book and count the rings: two long and two short for himself, three longs for Milton, four short and one long for Nitti, the Italian, three short for Costa and so on—the code of rings cast a net of shrill electric music miles wide over the forest. But now the age of innocence was gone.

Once the seed was planted there was no checking its growth. It was a fear that sent out fine roots and touched a nerve. A fantasy is like an extraordinary flowering tree, feeding on all the life in a body, branching and re-branching with endless elaboration.

The age of innocence was gone. He had been dragged naked from his solitude and privacy into public shame. He had been turned loose with the herd and grazed as they grazed.

The weeks in Rio were a confused memory to him now. He argued against it first of all. Lucy cannot be having a child. He became superstitious: 'If there is not a letter today, she *is* having a child.'

There were no letters. All these weeks he felt like a Gulliver pinned down to a mean, fine, stinging pain by thousands of Lilliputians. 'If she has a child I am trapped and chained to her.'

This was the whole origin of his fear laid bare. He wished to be chained to no woman. He wished to be alone. He counted the days when he would take the steamer to the north. The last few hours were agony, and when the coastal steamer put out to sea he could feel, amid the relief of escape from the land so thickly sown with his fantasies, an absurd fear that the steamer might be taking him back to England. His fear was a chameleon, he found, taking on all colours.

On the boat he avoided the passengers. He sat in the bar, drinking alone. Above all, he wanted to avoid the few English people. They, more easily than the Brazilians, would read his thoughts and suspect his words. There were two or three English business men and their wives on the boat, and on his last night when he went into the bar he found them sitting there drinking gin and already pink and drunk. One of the women, a fair-haired, tattered creature of forty who had been, she kept saying, a WAAC during the war, began to sing war songs. The others joined in. Tipperary, Keep the Home Fires Burning, Round the Corner, Behind the Tree—she went through the repertoire. Tears began to dribble down her cheeks. At the door the Brazilians looked in silently and rather afraid at the singing Valkyrie. Soon the words became coarser.

'Armentiers!' shouted one of the clerks, a lean, theatrical man with sad, ringed eyes:

'She took me out into a wood,
        Parlez vous,'

he shouted out.

'She done me all the good she could,
        Parlez vous,'

sang out the drunk lady. Some of the party lowered their eyes
uneasily: two of the men roared out with applause. The woman
laughed.

Johnson walked out of the hot bar into the air. They were
singing about him. He was a joke. He was a figure in the oldest
joke in the world. He was not alone, inviolable and exceptional.
He was a mere dirty comic character.

'It wasn't like that,' he protested.

He stood on the deck now in the soft night, watching the
white knives of water at the bows of the steamer, and wished for
one thing only: to escape from the ship. To get to the land, to
get up the river, to be far away in the interior of the country,
untouchable, to be purified by the torture of travel.

And then, in the ecstasy of this desire, the voice said:

'If there are no letters here, it's the last chance. Lucy is having
a child.'

There were no letters. He went down to the post office the
moment he had got his room at the hotel. Phillips came with
him. There were letters for Phillips but none for him.

They went back to the hotel and sat in the courtyard watching
the luggage arrive. Presently Johnson heard his voice say:

'I haven't heard from Lucy.'

'Good God,' Phillips said. 'Why not cable?'

'Oh no,' said Johnson. This simple idea had not occurred to
him.

'I should.'

'No, I think I'll wait.'

Cable and he would know! He would know that Lucy was *not*
going to have a child. And that would mean the deep guilt she
had awakened in him, would have to invent another fantasy.

He did not understand this or know it, except in an obscure, troubled way; he knew only there was for him a luxurious necessity in this self-torture. He clung to his illusion. The conscience of the puritan has need of its melodrama and mythology and he went up the river towards Wright, on the final stage, with the speechless fear of a son guiltily approaching his father.

CHAPTER FIVE

'IVE left him sleeping,' Charles Wright said. 'Nothing to worry about. It's just the country.'

They would simply have to sit there and wait. One evening they were sitting in the shed where their cases had been put. The place was overrun with rats and half-naked children. Seven of the children were Calcott's. He had a Brazilian wife. Two of them were negroid and two had coarse fair hair and blue eyes, but their skins were the colour of coffee. One of them could speak two or three words of English. Wright liked the children.

'He's all right now,' he said. 'I was afraid it was something bad. He'll be all right in a day or two.'

Wright got up and, squeezing Phillips' arm for a second, dropped his seriousness and his face cut into a deep smile of exclaiming happiness. 'Well,' he exclaimed, as much as to say, 'We are all here. That's fine. That's splendid.' Wright was a man with sudden boyish impulses of affection. The sun-burned skin of his hard skull gave the colour of a clay idol. He was like some small and diligent god. He was a busy god, strong, wiry and concentrated on his own mysteries, sure of his own intent, a little fey and overriding. A god whom one duped; an honourable man before whom one sinned in order to be equal with him? So Phillips speculated now he was alone with Wright. Wright had seemed like a trim and lazy stranger in his wife's house, a sober bird who has got in through the wire of an aviary of twittering repartee, queer feminine calls and counter-calls, smart, exotic, whistlings. During the war he had been a doctor in Egypt and Salonika. After that he had had some government job in India, but having some small means of his own, had thrown up that for exploration. He had been in Tibet and Mongolia, he had spent two years in the Antarctic and had come back with some reputation which grew after his South American travels.

A tall figure in dirty white trousers and jacket came swearing towards them. He was swearing at the children who ran to the house. At night, Calcott looked like some skeleton in its grave-clothes, prodding sardonically among the graves. By daylight his clothes wagged on his body like a flag round its pole. The lines and creases in his emaciated face were repeated with fantastic dreariness in his smeared clothes. His dirty blue eyes bulged out of their hollows with the brilliance of many fevers. He shaved only twice a week and at the side of his head the thick grey hair was bushed over his ears, but the rest of his head was streaked with lines of sallow baldness.

Calcott had spent thirty years in the country. He was a Cockney from Kentish Town who had made his way in the world, protecting himself from an acute sense of social inferiority by an unsteady and blustering contempt. He suspected every visiting Englishman of snubbing him. He had been snubbed and dropped in his time by what he called 'the shiny-arsed clerks' in the banks and head offices of the coast towns; there was some satisfaction for him in the fact that his marriage to a Brazilian woman damaged him in the eyes of these people. He liked to think he had damaged himself. He took it out of this woman when he heard that Englishmen were coming and after they had gone. He lived happily with his wife in the times between these English visits; but the sight of an Englishman upset him and then he set to and beat her. After these outbreaks he would shut himself up and pore over a Bible, not in repentance, but in wrathful conviction of the righteousness of his action.

Calcott's suspicions wore off if Englishmen stayed long enough. He was now at ease with Wright; but Phillips he looked upon as a new game for bluster and suspicion.

'You an Oxford and Cambridge man?' he said.

'No.'

'Reporter? Your boss said you were a reporter. A man who comes up here on a couple of weeks' picnic will find out damn all. The last one we had got one of those jiggers in his penis.'

And Calcott went on to a favourite theme: the insects that would get into their bladders, the fish that would tear them to shreds, the sting-rays which would cripple them, the boas that could crush them, the fevers they would catch and the lamp-mouthed alligators which lay waiting for years on the rivers for Englishmen with nice cultivated voices.

'You seem to have survived,' Phillips observed.

The contempt of Calcott was a crust easily broken. His response to this mild flattery was warm to the point of exaggeration. All sneering went. He pointed to the moon which hung like a huge Chinese lantern over the trees.

'For Gawd's sake look at that,' he said, with the pride of one who had arranged it all for them to see and now was awed by his handiwork. 'Makes you think of the old country, doesn't it?'

It was the one place the round yellow tropical moon did not suggest.

They went in to dinner.

A fan hummed in the room where they ate and a weak electric bulb put a dim dirty light on the white walls and the table-cloth. On the occasions when Calcott had English visitors, his Brazilian wife waited on them like a servant. He roared and swore at her, complaining about the food and sending her hither and thither. She was a short, corpulent woman with shining beetle-black hair, a muddle of negro and mongol in her features and her skin beaded with sweat. She was like a waddling spaniel expecting the whip. Every time she brought a dish into the room she had to push her way through her children who crowded watching in the doorway. Her voice when she spoke in the room was subdued.

This was a bad period for Mrs Calcott but she was indifferent. Outside in the courtyard, during the day, she squatted with her short legs wide apart and a baby in her arms sucking at her big breast which hung as heavily and passively and soft as another face; and she sang songs to the baby, fanning a charcoal fire with her free hand if a pot was stewing there at the same time. At times the song would stop short and her voice skirled out nasal and metallic to a servant or a child.

Calcott was in a nervous uncertain state on this evening. Phillips' politeness irritated him. The ease he had reached after having Wright for three weeks in his house had been destroyed once more by the arrival of the two younger men. He had once more to go through his repertory of bluster and self-pity. Wright himself would have come under his displeasure as a member of the household if he had not brought out the whisky at once. Calcott gave him a nod which could be translated, 'I wondered if I was going to have to remind you of that.' Calcott took the bottle, poured small measures into each glass and then put the bottle beside him, holding it for a long time by the neck.

'Your pal have some?' he said, nodding to the wall beyond which Johnson was sleeping. They said he would not.

'The more for us then,' said Calcott and helped himself to a third glass, this time keeping the bottle entirely to himself. But he began to mellow and conversation became easier. A look of aggressiveness which kept buckling into a smile of mixed amiability and suspicion came into his face, his lower lip dropped, showing teeth which had not seen a dentist since he was a young man.

He ignored Wright and directed himself to Phillips.

'Funny thing your pal coming up here,' he said.

'How do you mean?'

'He means,' said Wright, who knew what was coming from conversations he had had with Calcott in the past three weeks, 'he knew Harry's father when he was up here.'

'Did you? That must have been a long time ago. Harry's father's been dead fifteen years.'

Nothing could have delighted Calcott more than this inaccuracy. 'Seventeen years you mean,' said Calcott. And added ironically, 'So he's dead, is he?'

'One presumed he must be dead. He may not be.'

'That's better,' Calcott said.

Wright said tactfully, 'Mr Calcott was the last white man to see him.'

Calcott paused like an actor. 'He stayed in this house. He went out and that was the last anyone heard of him. You say he is dead. Well, you know better than me. But you can take it from me that if any man can go and hide in this bloody telephone exchange, it's a bloody miracle.'

60

'Telephone exchange?' asked Phillips.

'That's what niggers are,' said Calcott.

They knew the story so well, and they had been so pestered in England by reporters who wanted to know what they thought of 'the Johnson mystery,' that Wright cut Calcott short here.

'Just because Harry is with us, everyone jumps to the conclusion that we're going to find out what happened to his father. I've explained to Mr Calcott that this isn't so.'

Wright was sensitive on this point. He had made up his mind about 'the Johnson mystery' long ago and said that he was not going to waste his time on it. He had never quite got the suspicion out of his mind that Phillips, as a journalist, was in some way responsible for a newspaper article which had appeared before he left England, with the title, 'Another Search for Johnson. Son of Missionary leaves for Brazil.' Wright had his own plans. He had marked out his own square of country. He had seen it. The year before his marriage, when Indians had brought him down sick and nearly dying to the coast, he had stood on the edge of this untouched territory. He was too ill to cross to his promised land. He had lain at night in his hammock slung between trees by the river, looking across at it on the other bank, and forever he would remember the sun upon the wall of trees like the light on a woman's dress, the fantastic millinery of the tree-tops. There was a wild innocence, a feathery enticement and friendliness which were all the more irresistible because he knew all the treachery and chaos and suffering of immolation in the forest which were, by the most perfect art of the light, concealed. This memory, sharpened by his sickness, had always been photographed on his mind. Whenever he had looked at his maps afterwards and made his plans this artless scene came intimately to him. The earlier failure had haunted him. It was an irritating irrelevance that somewhere within a hundred miles of this place a missionary explorer had disappeared, almost staining a virgin country with the notoriety of his disappearance near it. Wright's passion was anonymity. He admired the missionary's courage in going into this country alone. He had studied every detail of the career of his competitor and had felt a professional pride in him. He had in England gone to see the missionary's widow and in that way had met her son. But some exclusiveness of the spirit was in Wright. He too had his private unknown land. He had seen its face and its dress. He longed to be in its body. The talk of the missionary's country and the mystery of his disappearance was talk of a rival and an attempt to enhance her attraction which he could admire as a connoisseur of discovery and adventure but which, now he was on the point of the heat of action, made him alert for any sign of betrayal.

But Calcott went on and Wright humoured him. The Johnson affair was the only thing in his life which had given him a vicarious importance. An eventless life, soured by loneliness and inferiority, had been given dramatic and even emotional point by it.

'It was seventeen years ago,' he lectured Phillips; and daring Wright to rob him of a new listener: 'I've got the cuttings. Some American paper sent a man to interview me.' He took out a greasy leather wallet from his pocket. Wright looked patiently at the worn-out, yellowing cuttings that were coming to pieces at the folds. He knew them. He smiled faintly at Phillips.

Calcott began to read:

'That Alexander Johnson, missing British missionary and explorer, was slain by cannibal Indians within a week of leaving...'

There was nothing in the cutting they did not know, though Calcott had added improvements of his own. '"He told me he had a presentiment that he would never come back," said Mr James Calcott, chief engineer...' Calcott put down the cuttings and poured out another glass of whisky.

'He knew he couldn't get through. He knew it. "You can't always back a winner, Calcott," he said. "It's like picking a woman," he said. "It's all in the hands of Gawdabove..." Those were his last words.'

And indeed so much time had passed since the missionary's death, and Calcott had so often brooded on the event and told the story, that the elder Johnson had become like a character in a bad novel to him, a character who has got inextricably confused with the character of the author.

But doubts evidently accused Calcott and doubts of the faith-creating kind, for he said:

'My conscience is clear anyway.'

He had nothing to answer up for. There was no fast train service 400 miles up the river. The Indians who lived in this bloody Turkish bath were not methodists!

'That was where he made his mistake,' Calcott said. 'There was no call for me to go up and see what had happened. I may be lousy,' he said, with a malevolent glance at Phillips, 'but I'm not an Oxford and Cambridge fool. If it's written, it's written. You can't alter that.'

'I see you are a fatalist,' said Phillips, taking the bottle from Calcott's side and helping himself, before Calcott had time even to glare at the outrage. He passed the bottle on to Wright. After an anxious glance he said ingratiatingly, 'What you say 'nother bottle, Doctor? The drinks are on you. We supply the grub, you bring your refreshment, eh!' Then to Phillips he said:

'If the twelve ruddy apostles had seen as many men die as I have, they'd have been ruddy fatalists. If it's coming, it's coming. You can't argue against it.'

You could not, they agreed. He nodded several times, staring into his glass. He muttered to himself, at first inaudibly; and then, more audibly, they heard him say, 'That's why I say it's funny your pal's come down here and it's funny he's sick. And it's bloody funny,' he ruminated more thickly, 'that it was just this time of the year his father spent his last night here.'

The remains of the meal were on the table, but talk began to break up, for Calcott was rambling on the borderland of an obsession with 'funniness'. It was funny they were here, the three of them; well, two of them and one of them next door; funny they should come. In a year they would be in England and he would be still here; that was funny. And there was something funny about everything. 'Where will we all be in a month's time?' he exclaimed, putting an arm on Phillips' shoulder and looking heavily, his jaw shaking, into his eyes. 'You don't know. I don't know . . .'

63

The voice thickened into ultimate incoherence and they left him. They saw him an hour later fast asleep with his head on a big Bible, open on the uncleared table, where the flies swarmed, and the bottle beside it.

Harry was left in Wright's room because it was the largest one. Wright and Phillips slept out in the courtyard. By day they took it in turns when he was not sleeping to sit with him. Calcott was often hanging about. If anything was wanted for Johnson he would be there, shouting orders to his family. Calcott's preoccupations with 'funniness' and his concern for one who was the son of the missionary led him to hanker after the privilege of sitting with Johnson. Once or twice in a day he would manage it and if found in the room would get up guiltily—because he knew he must not disturb the sick man—saying, 'Just having a word with your pal,' or, 'What do you make of his lordship now?' Then he would go off.

Phillips used to walk up and down the room making jokes and talking endlessly. He described the town. He described the canoes Wright had got. He was in an excited state, eager to go on. The illness of Johnson gave Phillips a kind of swaggering confidence; but he was entertaining. Once or twice he was on the point of mentioning Lucy, because Wright had said to him, 'Has Harry anything on his mind?' But he was too shy. He spent all his day walking restlessly about the town, sitting by the river or tramping to the outskirts until he had exhausted himself. He had no gift for being still.

When Wright came in he could find little to say. He looked at Harry and Harry looked at him. Each wondered what the other was thinking. They were reduced to the most conventional phrases. This delay in departure was worrying Wright because the time before the rains came grew shorter. But he concealed this concern from Harry.

None of them knew Harry's state and would not have credited it. They knew only that one moment he was better and, the next time they saw him, that he had strangely relapsed. Small things which they could not know affected him. But, on

the whole, he was much better and towards the end of the week
Wright came in and said:

'Look here, Harry, if I don't do something with Gilbert he'll
come out in boils or shoot up the town, so I'm taking him up the
river tomorrow. We're going after turtles' eggs and we'll have
to stay out the night. We'll be back the day after. That is, if
you're all right.'

'I'm all right,' Harry said. 'I'll get up.'

He was always protesting that nothing was the matter with him.

'You stay where you are,' Wright said. 'Get up tomorrow
afternoon for a bit if you like.'

But Harry himself understood then the crisis he had been
through. At the prospect of being left alone he was terrified. He
could have wept. All night he kept waking up listening for the
sound of their creaking hammocks in the courtyard.

But in the morning when they went he was astonished by his
elation. He was glad they had gone. He resented their going and
in the same breath felt a sudden freedom.

He lay alone in the heat of the day and the hours dawdled like
flies. He felt the fanatic gratitude of the sick that the world was
contracted to the area of the room and the white, excluding
walls. A lizard was in the room. He turned over and now it was
aslant as if gummed on the ceiling. He passed many half-hours
watching it, waiting to see it move.

He listened to the voices in the courtyard, the clatter of pots,
the gush of water, the singing and the quarrels. The shutters
were drawn. Until three there was only a strip of burning light
an inch wide from ceiling to floor and through this he could see
the thick leaves of a tree. They were like hot green tongues. He
got out of the hammock and stood still in the room, waiting to
see if anything would happen to him. He stood in his shirt. He
had got out several times to test himself. He knew his strength
was returning. Then he caught sight of his legs and he felt at
once: 'I am different.'

How different? He thought this out. Weaker, paler from
fever, no strength to go on? No. He could feel the return of
strength. There were two waves: the wave of feeling, 'I have

been plunged into fever and I am changed; when my strength is full again, it will be a different strength, something narrower and more intense. I have sloughed off something.' And the other wave was the guilt: the prisoner-in-the-dock-pallor of his legs.

He went to the shutters and peeped through at the sunlight. It was tigerish. The world was a gorgeous tiger and Wright and Phillips down on the river were like a pair of gleaming fleas in the fur. Johnson swayed when they came into his mind and his strength went. He put up his hand and leaned against the wall: the plaster crumbled. He had no connection with them. He smiled.

He got back to the hammock, waiting for his breath; but he was not as exhausted as he had been. The strength was there. He was building it up every minute. And he was aware that the change in himself was that the new strength was making a new self which had no knowledge of Wright and Phillips. There was the sick man who had known them and the new one who would not know them. He did not know how this would be or what it meant; he knew only that he was at the beginning of being another person.

The next time he got out of the hammock and was half-way to the window the door opened quietly and Mrs Calcott stood there. She had brought him some food. She stopped in amazement to see him standing there in his shirt.

'I'm looking for my clothes,' he murmured.

'I will ask my husband,' she said, 'when he comes back.' So it depended upon them whether he was to have his clothes?

She was thick-browed and swart, her lips moving rapidly and showing her white teeth, her voice small like a child's. When the amused light went from her she gazed at him with the lifelessness of the flesh. She was pregnant. He sat down on a chair and she put the food on a box by the window. Then she said:

'You speak good Brazilian. Mr Wright speaks good Brazilian. Mr Phillips—no.'

She laughed. She made statements about them all. Mr Phillips was fair, Mr Wright was short, her husband taller than Mr Phillips.

She stopped speaking and her arms fell to her side. When she saw he was looking at her swollen body, her face became softer, with a look of helpless tenderness. Her small lids half-closed as if she were glutted with the life that was swelling inside her and she made an instinctive languid movement of display as she let her arms fall back.

'You have children?' she asked him.

He started.

'I... No. I don't know,' he stammered. 'No.'

She looked very puzzled.

'I am not married.'

'That is not your wife?'

She pointed to a photograph which was on Wright's tin box in the corner of the room. Johnson looked and saw for the first time that there was a photograph of Lucy in the room. It was one Wright had brought with him.

Johnson looked at it with consternation.

'I didn't know that was there...' (He was saying the wrong · things and she must see it.) 'It is Mr Wright's daughter.'

She asked him whether he was betrothed to Mr Wright's daughter.

'No,' he said. He looked at her suspiciously. What did she know? Had there been letters? Had Wright told them? His look was hard and intense and the poor woman was as frightened as he was. Murmuring something she went out. He frowned at this small contretemps.

When she had gone, he ran his hand through his hair and stared at the door and then at the photograph. Then he got up and put the photograph under some books. That did not seem safe. Out it came and he held it helplessly, looking at it. The shock of seeing Lucy so different from the images of her that came to his mind arrested him. She was not like this, smiling, still, in a flowered dress, standing with her hat in her hand by the door of the house. This was not Lucy. But the house was real to him. There were the white pillars of the door and in the darkness of the hall behind her figure something glimmered. It was perhaps the light catching a vase on a chest or the suggestion of a door. There were Wright's things. Wright might have been at that moment inside the room with the door. There was the house; he had got into this house, he had talked in it. Afternoons swayed out of the past, like caravans, into his mind, evenings settled there. He marvelled at the separate existence of these things; and the effect of these actual sights, the white pillars, the shadow of a past afternoon on the step, even the mild pocked shadows on the brick, was to make his crime seem greater.

He could not bear to put the photograph back on the box, but when he caught sight of a jacket of Wright's hanging on the wall, he went over and touched it apprehensively. The cloth was curiously cold. There was a feeling of sacrilege which he had had once in touching a coat of his father's when he was a child. He put the photograph in one of the pockets.

All this had taken longer than he knew—days and weeks of England had emerged in these seconds—for Mrs Calcott came back and said:

'Why haven't you eaten?'

'I'm just going to,' he said.

She waited there. He had no alternative but to sit there in his shirt and eat while she watched him. He was too shy and confused to tell her to go.

Had there been a mail? he asked her. No, she said.

Then did she know, or didn't she? He could not bear her to stand there; but, just as he felt he would scream at her—not really scream aloud but feel his lungs and his throat make the movements of a scream—just then, she scraped out of the room.

'Ask for letters,' he heard her say to some woman outside. 'Ought not to be out of bed. Has a white skin like my husband. Is very young.'

He pushed angrily at the box, pulled down Wright's coat, looking for clothes, to escape the humiliation of standing in his shirt only in front of this woman who was going to have a child. A lunatic must feel like this, with Calcott's wife for an attendant.

'I'm not going out of my mind,' he said within himself as if his head were a room in which he was sitting and talking with two or three other selves: the pre-Lucy Johnson, the Johnson

68

sickening on the river, and the new, hardening one now getting back strength. 'They thought I'd lost my nerve so they took my clothes away.'

He thought of the Calcotts as gaolers.

Late in the afternoon when this crisis had passed, Calcott himself came in to see him. This was Calcott's marvellous opportunity. They had never had much conversation. Calcott was tired of Wright and still took offence at the manner of Phillips. He had a sentimental interest in Johnson. Calcott had been longing to convey this to him.

'So your pals have left you,' he said.

'Yes.'

'Mind that?'

'No.'

Calcott took off his straw hat which left a red band on the grey sweat of his forehead. He sat down on the box near the hammock and said in a low voice which was really respectful, but which sounded because of his nervousness like the sniggering beginning of a dirty story:

'I knew your father.'

'So I heard,' Johnson said.

'Who told you?' asked Calcott suspiciously.

'I expect it was Charles Wright.'

'Oh,' said Calcott in a disappointed voice.

They sat in silence.

Calcott smiled. When this unfortunate man smiled, though (as at present) nothing but a simple and innocent warmth inspired him, his face could achieve nothing better than the unconvincing smirk a prisoner might direct at a judge when pleading for leniency.

'He was all right,' said Calcott confidentially. 'Your father was all right. More than you can say about most missionaries, but when you get a good 'un, they're good.'

Johnson was sensitive about his father's profession.

'I was a boy when he died,' he said.

'What I said to Wright,' said Calcott triumphantly.

'Your pa,' continued Calcott, 'stayed here with me. In this

room. This was the last room he ever slept in. The last room, I mean to say, in a house. What do you make of that?'

Johnson put his book down and looked at the room. It was not news to him that his father had stayed at the house. He could see no sentiment or drama, such as Calcott saw, in the fact.

'I don't remember him much.'

'He changed my life,' said Calcott. 'Here . . .' he said, getting up and going to the barred window. A woman was carrying a petrol tin of water across the courtyard. 'Go and get my Bible,' he shouted at her in Portuguese. It was comic Cockney Portuguese.

She went across the yard with the tin, mumbling, and presently the cross-eyed child brought in the book and gazed at Johnson.

'Clear out,' shouted Calcott in English, raising the book at the child. On bare feet the child skipped out of the way and closed the door.

'That's what your pa gave me,' he said, handing the book to Johnson. Calcott moved to a low chair with his long bony legs sticking out like the legs of a spider. 'He gave me that when he married me and Mrs C.'

Calcott got up and went to the bars of the window to spit. He grunted as he came back to the chair. He hesitated before sitting down and then decided to move the chair to the window in order to be near for spitting again.

'There was a proper blow-up with the priest.' Calcott lowered his voice: 'They're all Romans. On top, I mean. Underneath Gawdabove knows. The old bastard said she was damned. I told him if I saw him speaking to her again I'd throw him in the river. They've got a nerve though, those Romans. He said to me, "You wife's living in adultery." Bloody insolence, talking like that to a man about his wife.'

'Anyway,' continued Calcott, '"You've missed the bus," I told him; "she may have been living in adultery and no credit to you, up to yesterday, but she isn't today, see?"'

'What did he do?'

'What could he do? Why didn't he go and clear up the town.

Women's the only industry on this river now rubber's gone up the spout. Always was the only industry. That's what he did for me, your pa. "If we've sinned," he said, "we've sinned, but we can put it right. Don't be one of the crowd that leaves it till Judgement. You don't know when that will be," he said. "It's not your fault, Calcott," he said. "You're more sinned against than sinning." Then he did us. Then he christened all the kids. There were three then.'

It appeared, however, that the nature of woman was deceitful, as Calcott continued his life story. When the other children arrived his wife took them one by one to be baptized by 'the Romans,' on the quiet.

'They're all rotten, the people of this country,' Calcott said.

'My father seems to have caused you trouble,' Johnson said.

'You say that!' exclaimed Calcott. 'You're young. You don't know what you're talking about. Excuse me saying so, no offence, but you don't know what your pa saved me from. Here—supposing you had a woman—I mean to say we're all *men*, aren't we?—well, and you got her in the family way; well, you can talk all round it and up and down it, but you know you've done wrong, don't you? You've got a conscience. Gawdabove sees everything, you can't deny that. You know you'll have to answer up.

'It may be the woman's fault,' said Calcott, 'but *you*'ll have to answer up.

'And if you was my son and you done that,' said Calcott, wagging his finger emphatically, 'I'd say to you, "It may be the climate but you're a dirty bastard all the same." And so would your pa.'

'Missionaries in our part of the country aren't so particular,' Johnson said.

'You're telling me,' said Calcott. 'But your pa was different. The day before he came here he had been up to a little place eighty mile up the river. There was a man up there, an Irishman called Macguire or Macguinness, I've forgotten his name, and anyway he's dead now, used to be a Roman and then came over. He'd been up there twenty-five years. No one ever went to see him. Result, as per usual, he goes off the straight and narrer. A man of fifty. He takes on an Indian woman and then there's a jolly little nest of them, all of them happy as tomtits, you bet. When anyone comes up from the mission he gets word of it, buries all the bottles, pins up the texts, and they come ashore and find him in the middle of a service. He'd got it all worked out to a "t." That was all right. But being Irish he gets beyond himself. Starts giving out he's a pope or witch-doctor, worshipping the devil or something he's got hidden away in the woods, like the Indians. Your pa heard of it and copped him at it, wearing no clothes and covered in paint and feathers, doing a bit of a hop, skip and a jump round a blooming bonfire with the rest of the boys and girls. Proper Irish. He gave the ol' man a hiding. "Calcott," he said, "I drove him down to the river with my umbrella and I made him scrub the stuff off him." Your pa wasn't having any. He was a gentleman.'

Johnson laughed. Calcott started and then laughed too. He gave a huge spit out of the window and laughed hard. He was glad to have made Johnson laugh.

'Cor!' laughed Calcott. 'You're right. He was a cure, your pa!'

Calcott's face became liver-coloured. Each laugh began with a snigger, ran on into a cackle, barked out full and ended in a sweeping throat clearance. Tears were in his eyes. Johnson laughed quietly.

'Have a wet?' said Calcott. Then superstitiously, 'No, better not.'

The pleasure he had in making Johnson laugh made the suddenness of his desire for a drink seem like one of those inspirations which all drinkers wait for, with a mind almost devout. He left the room and soon he was shouting for Wright's whisky and pouring out a glass for himself. He drank it. Then one for Johnson in the same glass. He drank that. Well, another for himself and another wouldn't hurt Johnson. He drank two more. He sat down alone in the other room, drinking first his own glass and then Johnson's. There was only one glass.

Then he felt guilty that he had not behaved with the decorum he was sure Johnson and his friends would have observed. So he poured out another glass and brought it into Johnson's room. Making a ponderous gesture and holding his lean chin high he said, 'Mr Johnson.'

And he raised his glass. Johnson waved a hand. Calcott bowed. But the bow went too far forward and he was obliged to sit down quickly in the chair again.

'Funny thing,' he said.

'Damn funny thing—you ask me—you and your dad,' he said.

Then he fixed Johnson with a warning eye.

'Your ol' dad—not human. Never touched it.'

'What?'

Calcott raised an empty glass despondently.

'Never touched it. That's what was wrong with him. Wasn't human. I'm human. That's what's the matter with me. Always have been. Too human, you might say.'

He stuck out his bony jaw in a brave attempt to repel frail humanity, but his face buckled into its whisky lines, and an expression vulture-like but sentimental came on it.

'Know what them dago kids call me? Uncle Jim! They do. Straight. Uncle Jim. I give 'em things sometimes, the poor bloody little devils.'

CHAPTER SIX

A STORM broke over the town in the night and was prolonged until within an hour of sunrise. First of all a small wind ran loose in the town and the black water of the anchorage; then it whirled and whipped until it was like a howling tribe, slamming doors, blowing off tiles and streaming through the trees. The forest was like a sea in tumult and the town like some gimcrack raft plunging about on it. But no human sounds came; the people slept, or gaped at their theatrical climate. Then the thunder went off like guns and bumped about the roofs and the rain fell—not a thin sloping rain but a rain in vertical solid rods. The rain turned the air into another forest, hissing and impenetrable.

The rain caught the Calcott family in their hammocks. Out they jumped and ran into the house, shouting at each other. Things fell over and smashed. In the middle of the storm there was a terrified scream from the parrot in the courtyard. Johnson got out of his hammock to close the shutter and saw a river of water swirling in the courtyard and flowing into one of the doors.

The storm passed and in these pauses one heard the water streaming from the roofs; but in half an hour it was back again making violet lakes of lightning around the town and the thunder went off once more. Then the spectacle died away into hour after hour of fitful lightning—as if someone were playing with electric switches.

In the morning Calcott himself brought Johnson's coffee.

'Did the other two get it last night, do you think?' Johnson asked.

'They copped it all right, you bet,' said Calcott, grinning. 'I've been damn nearly drowned up there meself before now.' he added: 'The parrot's finished. Drowned. Hear it?'

Calcott had gone out in the night to rescue the parrot. In vain. 'Don't like it,' Calcott said in his most lugubrious voice. 'Bad sign. Means something.'

He sat moodily in the room watching Johnson. He looked as though he wanted to stay near Johnson all day.

'I wouldn't have lost that parrot for a thousand pounds. I've had it since your father was here,' Calcott said.

Nothing Johnson said could cheer up Calcott.

'She doesn't care,' Calcott said, meaning his wife. 'None of these dagoes care. They'd sit and watch an animal die. She would. No feeling.'

He stood up on his bent stick legs and looked moodily at the room.

'I'll get Silva to come and see us this afternoon,' he said gloomily, as if a situation so melancholy demanded Silva, as a corpse demands an undertaker.

'He's at the works, isn't he?' said Johnson politely.

'Yes. You know—Shakespeare. Got me? Educated bloke. I'll get Silva in.'

'If you would tell me where they've put my clothes,' said Johnson, 'I'll get up.'

'No need to, ol' man,' said Calcott. 'I told Silva you was ill. He won't mind.'

But Calcott was eager to do anything for Johnson. He went out of the room and returned with the clothes. He put them down expecting Johnson to get up and dress at once. Johnson did not move. Calcott was lost for words and walked about the room. He gave Johnson a look full of significance. Significant of what? Only Calcott knew.

Men who live cut off from their own race, in a foreign country, fall back for support on a peculiar inner life of their own. They have to make something with which they can resist the country, and Calcott had managed to reserve in his exotic surroundings, and in spite of his Brazilian wife and his half breed children, an inner temple full of holy relics of life as it might be lived in the Old Kent Road. There was the Union Jack spread over the Brazilian piano, and a book of English ballads, never opened but always on it. There were pictures of boxers and actresses cut out of illustrated papers and pinned to the wall. There was the linoleum in his own room, a coal-scuttle and fire-irons which his mother had sent out to him years before. The old lady had never been able to understand that a fire was not lit in these parts from one year's end to the other; but Calcott would not throw these useful things away. He kept them like sacred vessels and implements. And Calcott's resources were not external only. He had others. It was a sign of the attachment he felt towards Johnson, an attachment which was fervent now his companions were away, that he was about to reveal his great resource, the extraordinary mythology he had created. The only difficulty was that Calcott, in his solitude, had got into habits of allusiveness in conversation which were full of meaning to himself but which meant nothing to others. He imagined himself a high priest of insinuation and diplomacy. When Calcott said, 'We'd better get Silva in,' he wished to convey that Johnson, albeit unknowing, had provided him with an opportunity of exploring heights of 'funniness' which he (Johnson) would not credit.

'Silva's an educated man.' Calcott explained again.

The day passed more agreeably than the earlier ones. Harry got up and thought of going down to the river, but the town swayed in his eyes and he stayed under the trees by the house and few people noticed him. He felt he was not in control of himself, that his body might be still lying in the hammock like a chrysalis suffering some change which he could not command or alter. He had looked at himself in a mirror and had indeed shaved off his days of beard, and he had been surprised to recognise no radical change in his appearance. But he knew that in the last few days a change had begun in him. He had passed a crisis. He had begun a new road.

In the afternoon he went back to his room and sat deeply preoccupied. Mainly he was thinking now not of Lucy and her stepfather, but of the journey before him. He was thinking of the country beyond the river and of the weeks ahead, but in terms of paddling, mileage, hours of distance, thickness of scrub. He did not think of Wright and Phillips as being with him. His thoughts were visually close and vivid, the pictures not merely seen but clothing him as one is clothed by the warm drugged pictures of a dream. He sat on the small straight chair through the blue swamping heat of the afternoon while the household slept the siesta. Even the flies were still and the cold lizard on the ceiling did not move from its dark corner. He was surprised when voices broke into the house and Calcott came into the room with Silva.

'This is Mr Silva,' said Calcott.

'Ah!' exclaimed the Portuguese, and increased his flow of excellent English. His long acquaintance with Calcott had, however, damaged an almost too perfect Oxford accent. He was wearing a violet-striped pyjama coat in honour of meeting Johnson.

'Maria,' shouted Calcott from the door. 'Bring the table.'

'You know the table?' asked Silva.

'Silver'—Calcott always called his friend by this name— 'Silver's a dabster,' interrupted Calcott. 'We get something every time.'

'Ah, no,' said Silva modestly, lowering his eyes. 'Mr Calcott is a hell of a medium,' he said, with polite detachment.

One of the women brought in the table. It then dawned on Johnson that Clacott and Silva were going to conduct a spiritualist seance.

Silva sat back. A cigar was between the fingers of his small, white, right hand. Silva worked under Calcott at the Power Station. He was a fat man in miniature, miniature in blandness, like a monkey of a very old civilisation. The cheeks in his small head were chubby and darkened, like the smoke-deposit on the bulge of a lamp glass, with a day's growth of beard. A black voice, enormously deep, came from the full lips of this very small man. Once a clerk in an English shipping office in Lisbon, Silva had early in his life decided always to attach himself to the English. He read English books. Intelligent, imitative and simple hearted, his admiration for the English sprang from the observation that they had a lot of money and were, by his frugal standards, very careless about it.

Silva was a man of talents. He had to perfection the art of living on very little money. He had the art of being inquisitive without being offensive. He had the art of falling on his feet wherever he went, swinging from one job to another, from one country to another, like a monkey from branch to branch in a world that was a luxuriant tree. He knew where the best women were in every town he visited. In a small worn notebook he kept the names and addresses of women he had known in London when he had gone there as a tutor to the son of a Scotch Baronet who had Portuguese interests. Silva had been a waiter, too. He was without prejudices. He had the art of avoiding trouble, the art of pleasing everyone in the way they wished to be pleased. He was as discreet as a cat.

Silva did not believe in God, but this did not alienate him from the people who did; on the contrary, it attracted his interest. Similarly he did not believe in spirits; so Silva willingly became vicariously English in his seances with Calcott. And Silva believed in doing things habitually and well.

One day Calcott and he discovered that a spirit who gave his name as 'Hamlet' wished to communicate. Silva explained that this must surely be Shakespeare's Hamlet. Calcott was impressed. 'To be or not to be ... you mean,' he asked Silva nervously. 'That is the question,' said Silva. 'Yes, that is the one. Let's ask him.' The spirit *was* Shakespeare's Hamlet. 'You ask him something,' said Calcott, out of his depth. 'What are you doing Hamlet?' Silva asked the spirit. 'Dancing,' the spirit replied. 'Beautiful music and girls.' And then he broke into poetry and said he would drown the Mayor of Lisbon. Mentioning Shakespeare, the spirit said he didn't like him. He was difficult. The Duke of Wellington and Caruso were with him. They were eating. What? The table became very excited and this point and leapt about. At last it quietened down. 'Steak and onions,' was the reply. Calcott's subconscious mind, with its nostalgic memories of the Old Kent Road, at last awoke to the inventive genius of Silva and, rebelling against Hamlet's stress on moonlight and girls, began to make its mark.

They took it in turns to type out the messages. Now a thick bundle of MS. had accumulated, full of the eloquence of this diffuse character.

Hamlet became the creation of their combined but very different geniuses, the outlet for the fantasies of Silva and spleen of Calcott. Some days he was laconic. 'Fed up,' he said. Or, 'Shakespeare's drunk again. I'll kill him.' On other days he was rhapsodising about a Portuguese lady called Teresa who had yellow eyes like the Amazon. 'Have to get my pay in advance.' Teresa faded. 'Dago got her. No whisky. Shakespeare drunk it.' Followed by 'British Navy very strong.'

This last episode corresponded to a period in which Calcott had been the unsuccessful rival of Silva for an Indian girl in the town. Silva, ever tactful, ever the artist, had contributed the last sentence. Passions which might have become dangerous were thus sublimated in this strange work of art.

'Well, Silver, ol' man, what about having a try? It mightn't do much today, a stranger being present. But,' said Calcott, raising his eyes to the ceiling and speaking louder so as to be heard clearly in the beyond, 'we've got the son of the Revd Johnson here who passed over, and he's very interested.'

'Perhaps Mr Johnson would like to try,' said Silva.

'Oh, the spirits won't care for me,' laughed Johnson.

'Not if you're hostile,' said Silva.

Calcott anxiously watched the effect of his proposal upon Johnson. He conveyed that he had reserved the honour and interest of this most private excitement for Johnson alone; that he was admitting him to an innermost secret. And Calcott also rolled his big faded eyes significantly, indicating that this seance might have a far greater importance than Johnson at present supposed. It would be sinister and significant because Calcott had the gravest interest in Johnson's life. He was doing Johnson a good turn.

'You watch Silver,' said Calcott. 'He comes in every Thursday and we don't always click, but we usually do. Don't we, Silver, old man?'

'You smoke a cigar?' said Silva, taking three cigars from his breast pocket.

'Mr Johnson,' said Calcott, leaning with command over their heads, 'have one of Silver's cigars.'

Calcott sat down at the table and then Silva sat down.

'Don't close the shutters yet,' said Calcott. 'Have a look at this.'

Now they were ready Calcott prepared an impressive delay in the proceedings.

He opened the pile of typescript. Typed in a faint violet ribbon in single spacing, thumbed, smudged, inked with corrections, were the verbatim reports of the seances of many years. Calcott turned over the pages in loving reminiscence, applauding, speculating.

He read out many passages of Hamlet's words from the other world.

So Johnson joined them. He put his fingers on the table. They all waited in silence. The only sound was the clipped puff of their cigars. Calcott looked up at the ceiling. His face was tense with expectation. He fervently believed in the existence and voices of the spirits. Politely, after Calcott had raised his big eyes to the ceiling, Silva raised his small ones with an abstracted blink. The air of the room was close.

'You have to wait sometimes,' said Calcott, without looking down from the ceiling.

'They're sleeping,' said Silva, in a low gurgling voice to Johnson. They waited for many minutes.

'Funny,' said Calcott. He gave a look under the table. 'Thought someone might have their foot on the leg.'

He looked up to the ceiling again.

The day after tomorrow, Johnson was thinking, Wright and Phillips will be back. He was in the middle of this thought when he felt the table give a jerk at Silva's end.

'Ah!' exclaimed Calcott.

'Who's there?' he called.

The table did not move.

'They're coming, said Calcott, now intently listening.

'Shall we ask if there is anyone who wishes to say something to us?' said Silva cermoniously to Johnson.

'This is your show,' said Johnson. 'Don't ask me.'

'Come on,' said Calcott. 'Speak up, Silver.'

He was always a little nervous of addressing the spirits himself. So Silva asked the question in his most polite manner.

'One knock for yes,' he reminded them. 'Two for no.'

The table made two clear movements.

'Mucking about with us, eh?' said Calcott.

They all began to get restless. Their arms ached. Their wrists relaxed on to the table. Calcott was the first to take his hands away.

'They are sleeping just as I said,' Silva smiled confidingly.

Johnson smiled at Silva's remark and Silva smiled back as if a wick had been lit and turned up in his head, a smoky smile. All peoples regard the man who is foreign to them as mysterious, a creature of inexplicable ways and occult sympathies. Silva perceived by a certain restraint, nervousness and deference in Calcott's attitude to Johnson that the latter was, in some way, more privileged. The peculiar mystery which the English understood was the mystery of going into strange lands and making money. Johnson's father had been a missionary, for example; missionaries always had money, English money. Silva's ingenuous interest was roused. Dreamily he looked at the young Englishman, who was clever enough to pretend like himself that Calcott's spirits were real, and thought of the probable gold mines that were hidden in the country to which the expedition was going. To Silva, one had only to follow an Englishman and, where at last he stopped, there would be gold. Or if not gold some valuable equivalent. While they sat round the table and Calcott talked of the utterances and sprightly behaviour of the spirit, Hamlet, Silva gazed at Johnson's shaggy black hair and finely lidded eyes, and thought about this.

And Johnson's thoughts about Silva ran on similar lines. Calcott was a joke, his seances were a pathetic parlour game. Johnson had met many lonely and uprooted Englishmen of the Calcott type who between their cups sank into fanatical and moody obsessions about the end of the world, the abominations of the Pope, the literal interpretations of the scriptures or the prophecies on the Pyramids. But the same obsessions transferred to a Portuguese were another matter. They were still quaint but

81

translated into another tongue they had a hint of mystery. When Silva said, 'They are sleeping,' the words seemed to Johnson fanciful and, still, in their fanciful way, haunting. Since he was essentially a practical man, it was the humour of the words that haunted him. The way they established themselves in his mind was a joke to be remembered.

The shutters had been opened again and they were all sitting in the light. Calcott was talking. He was boasting to Silva of the old days and how he had met Johnson's father and showed off in front of Johnson by speculating about his father's fate.

'This,' he repeated the story, 'was the last room he stayed in.'

The dramatic significance which Calcott's Cockney imagination saw in this fact, was lost on Silva, but his quick curiosity noted that it seemed to mean something to the mysteriously privileged race of gold-knowing people called Englishmen.

'What d'you say to another try?' said Calcott with embarrassment after a while. 'They were there all right.' He looked in a baffled way into the middle air of the room. 'There was a bloke mucking us about. Might have been Hamlet. What do you say?'

Johnson asked to be left out of it this time. And Calcott approved of this. Silva agreed with Johnson. The shutters were closed by Johnson and, putting down the stumps of their cigars, the two men began the rite again.

It was evident that now the experts were unhampered there would be results.

'See that,' murmured Calcott significantly.

'Hell,' came the deep, respectful voice of Silva.

The table began to sway and tremble under their fingers.

'Is there anyone there wishes to speak? If so give one knock for "yes," two for "no,"' Calcott said, raising his eyes to the ceiling, and before he finished the table gave a strong affirmative movement. 'Who are you?' The table leapt again into action. Calcott clutched it as if it were running away from him.

'L,' cried Calcott, head on one side as if listening to an actual voice. 'E.S.L.I.E. Leslie, yes! My brother!' exclaimed Calcott. 'I was telling you about him the other day.'

'Ask him something,' said Silva.

'Is mother there?' asked Calcott.

'Yes,' the table answered.

'Bring her along.'

'She's busy.'

'What's she doing?'

'Too much work.'

The table laboured at its replies.

'I'd like to speak to her.'

'Shut up.'

'That's not the way to speak, Leslie,' said Calcott in an injured voice.

'I've got Mr Silver and Mr Johnson here.'

'Bovril,' replied the spirit.

'Hell,' said Mr Silva.

'Hungry,' said Johnson.

'Are you hungry, Leslie?' asked Calcott, distracted by Johnson's amusement and Silva's comment.

The spirit reverted to an old theme of these meetings. 'Hamlet's drunk.'

But no, the spirit said, Hamlet was not there. How did he know then Hamlet was drunk? Answer: always drunk. Father's ghost.

'But Hamlet's a ghost too. It oughtn't to worry him now,' said Johnson, interrupting. 'Ghost doesn't worry ghost.'

The table paused while the three discussed this point. Calcott was very struck by it. 'Can a ghost see a ghost, you mean?' said Calcott.

'He wouldn't know it was a ghost, would he?' pursued Johnson, now thoroughly enjoying himself.

'That's right,' said Calcott. 'What do you say, Silver, ol' man?'

'I think everything is possible in the world of spirits,' said Silva very gravely in his deep voice.

'Leslie,' called Calcott, 'are you telling lies?'

The table leapt and hit Calcott hard in the stomach twice.

'You see,' said Silva.

'Al right,' said Calcott. 'We only wanted to make sure because

Mr Johnson is here.'

'No,' replied the spirit.

'Oh yes, I am,' called Johnson from the chair.

'No, no, no,' persisted the spirit.

'Funny,' said Calcott. The table was rocking swiftly in action. Back and forwards it went between the two. Calcott's shirt-sleeve came down and he rolled it up. The two men began to sweat under the energy of the table.

'He is not here,' the spirit said.

'Of course I'm not,' mocked Johnson. Calcott glared at him.

'Leave this to me,' Calcott said.

'Perhaps,' said Silva, with a mild detached delicacy, 'he is speaking about the other Mr Johnson. The father,' he said, 'of our friend.'

A sudden embarrassment overcame Calcott. He had had this at the back of his mind when he proposed the seance in the first place; that, in some way, they should honour and impress young Johnson by putting him in touch with his dead father. The deference Calcott felt to the young man made him hesitate in openly saying this and he had not known how to broach the matter, and the confusion of the seance so far had not helped him. But now Silva had done the task for him Calcott broke into a sweat of relief; in the semi-darkness of the room he drew hard on his cigar stump.

'That's funny you say that,' Calcott said in a blatantly unnatural voice. 'I was thinking the same myself. Do you think we ought to find out, Silver? That's what he's up to all the time, trying to get through to our friend.'

'It depends on what Mr Johnson thinks.'

'Me?' said young Johnson. His amusement had stopped. He was embarrassed at being the centre of interest and he had no wish to involve himself in what seemed to him a bad joke once it brought his father into it. The idea of a faked communication with his dead father irritated him. But, he was above all a tolerant man, and moreover he was honest enough to note a curious, morbid interest in what would happen.

'All right,' he said.

Calcott made a dramatic gesture.

'You never know,' apologised Calcott. 'He might be some help in your job.' But a sudden alarm seemed to seize Calcott, for he presently asked Silva to speak. 'Silva's better than me,' said Calcott humbly. 'He's an educated man like your father was.'

The English class system survives to the very brink of the next world.

So Silva, in his most ceremonious manner, asked the spirit if he would have the delicacy to explain for the benefit of their great friend the significance of his words.

'One knock for yes. Two for no,' interpolated Calcott, not quite able to keep out of this crucial drama.

The spirit repeated, 'Johnson not here.'

'Ask him where he is.'

'Would you have the kindness to tell us where he is?'

There was no coherent reply to this. The social approaches having been made, Calcott recovered his old peremptoriness.

'Can you see him?' he called into the air.

'Gold mines,' repeated the spirit.

'Can you see him!!'

'No.'

'Is he anywhere there?'

'No.'

'Is he alive?' Calcott's voice went high on the question.

'Gold mines.'

'Ask him what gold mines,' said Silva.

'What gold mines?'

'Yes,' said the spirit.

'You mean that Mr Johnson's father is not dead but has found a gold mine in the forest,' said Silva in a dispassionate voice.

'Wash out,' said the spirit.

A shout of laughter went up from Johnson and one of his rare frowns of cold annoyance, the contempt of the insulted artist, passed over Silva's invisible face. Johnson was on his feet saying ironically:

'Well, I'm glad he's not dead and that he's struck gold. But he's been keeping pretty quiet about it.'

He opened the shutters and stared out into the courtyard and the black sky where the big stars hung like white fruit. He had no idea that he was trying to break up the meeting. And when he saw the stars and heard the women singing in the kitchen, his genial thoughts vanished and with a feeling of cold revolt he thought of Lucy again. He saw her head and her shoulders in the darkness.

Calcott said, from his chair:

'I don't believe your father is dead. No one's ever seen him. I've always said to Silver, "That man's as alive as you and me, Silver," haven't I? The message is that your father is alive.'

Calcott stood up, feeling the dark wings of the drama he had created standing tall and ever growing beside him.

'Oh yes,' Silva replied. 'Maybe he's alive.'

'He *is* alive,' declared Calcott. 'Mark my words. I wouldn't mind betting anything I've got on it. You don't believe it, I know. But why should he say he was alive if he wasn't? Wouldn't do him no good.'

Calcott was inclined to resent Johnson's ingratitude for this immense effort of research which he and Silva had staged. He suspected at once that Johnson's attitude was due to class feeling. He felt snubbed. That night he showed Johnson his feeling by a score of sneers at 'Oxford and Cambridge men' who came out to 'this country and think they can swank through it like the Bois de Bolong.'

'What *is* education?' he said in a more friendly tone. 'You're educated. Your father was educated. Silva's educated. Well, what of it? Eh?'

Calcott could not leave Johnson alone. He was completely fascinated by the fact that Johnson was the son of his father, whom he (Calcott) had known. He drove his own children out of the way, shouting and screaming at them for a lot of filthy niggers, saying, 'I've made a mess of my life, I have. Women have been the ruin of me. Keep clear of women. They get you if you're not careful. Take my advice and never marry a dago. All women are dagoes.'

Silva took his leave very late and walked away like a small cat.

He went in and out of half a dozen bars, standing against the counter in the smoke, entering without formality into every argument that occurred, improving it and speaking so that soon the conversation was his. He could assume possession, in his peculiar disinterested way, of any conversation. Unobtrusively he spread the rumour that the Englishmen were going up the river for gold.

CHAPTER SEVEN

THEY were away and his strength was returning. Harry walked into the town. He sat watching the Indians sailing their canoes in the morning squalls. On the bank the flies came in clouds and pestered, but on the river there would be no flies. He had no mind now, it seemed; his mind had been poured away in his sickness. He had only a body that could be moved about from place to place, enjoy the lick of the air upon its skin and suffer the bites of the flies.

When he got back to the house he went to look at his canoe. He unpacked it and put it together. They were going to try out this canvas canoe on the river. It was a special fancy of his to try this boat because it would be swift and would draw little when they went up drying streams in the forests. At the rapids it would be portable. He had pictured himself going out from the camp in the evenings to shoot game on one of those lagoons that were left behind in the forest when the waters fell in the dry season. He was deeply engaged when Silva appeared in the shed.

'You want me to help you take it to the river?' said Silva.

'No,' said Harry. Then he looked up and said, 'Perhaps later on.'

Silva stood watching. Presently he began to say one or two things about the country. Johnson grunted.

'I know the place you're going,' Silva said.

'Oh, *do* you?' said Harry, not believing him.

'Yes, I know. I have been there,' he said.

'When?' Johnson asked, now interested.

'Long time ago,' said Silva, 'before I came here.'

Harry looked at this figure. He could not imagine this fat little man on any expedition.

'It must have been,' he said.

'I went with Mr Wright,' Silva said.

Johnson put down his knife and looked at Silva. He tested him with a few names.

'Yes, I've been there,' Silva said mildly. 'Mr Wright was sent back sick—like you—he quarrelled.'

Johnson had heard this story of Wright's quarrel. It was Wright's earlier expedition. He had joined a party of gold prospectors—or so they called themselves—but after weeks in the forest they began to fall out about direction. The man who had put up most of the money could not stand the hardships and wanted to go back. Wright had stayed, but a few weeks later himself was brought back ill. He had had a poisoned foot. But he had seen at that time country to which he burned to return. Silva repeated the story.

'I want to come with you,' Silva said simply.

'You had better ask Wright about that,' Johnson said.

'You introduce me to him,' Silva said.

'But you know him. Haven't you spoken to him?'

'Yes, I know him—but you introduce me to him.'

He paused and added modestly:

'I know all this country.'

'So,' laughed Johnson, 'that's why you talked about gold last night.'

'*I* talked?'

'Well, the spirits.'

'Oh, the spirits.' Silva made a disparaging gesture. 'You don't believe in the spirits, you don't believe they talked about your father?'

'Of course not. Do you?'

'Oh no,' said Silva.

'But Mr Calcott believes.'

'Yes,' said Silva, still in his simple way. 'Mr Calcott is very lonely.'

'Gold or no gold, you will take me?' Silva said after a while.

He spoke with an eagerness shadowed by a world of resignation and nonchalance. He dug the toe of an old cracked shoe in the dust making artless circles.

This was a bad place, Silva said, for a man like him. There were too many women. He adored women but not too many— not every night. There were the spirits two nights a week at

Calcott's house and then the rest—women. It was too much.

'And I am tired of this place,' he said. 'I would like to help you with the boat.'

Wright and Phillips were to return in the evening. Harry pushed the fact from his mind.

Before noon a launch from the coast arrived with the mail and Calcott collected the Englishmen's letters. He had them in his hand when he met Johnson in the street.

'Here you are,' said Calcott.

One by one, avidly reading the addresses of English correspondence—for he had not had a letter from England for years—he handed over the letters. Johnson took them and stood in the street going through them. His heart was beating loudly. Now he would know. But he had gone past knowing. It seemed to him that the whole question had become past, an academic matter concerning another man—one who had tramped the streets of Rio in agony and had sat in the slow suffocation of oncoming sickness before the unrevealing forest screen.

There was a letter from his mother. There were letters from his brothers. There were several letters in unknown hands to Phillips, one from a newspaper. (Yes, what was Gilbert going to write in the papers? Were he and Wright going to be in it?) There were letters for Wright, two from Mrs Wright to her husband. But there was nothing from Lucy. Nothing at all. Unless there were letters from her enclosed in Mrs Wright's envelopes. He looked at them again and felt them.

Slowly he stuffed them in his pocket and walked in silence up the street with Calcott. It was deserted at this hour. The shutters closed, nothing but hens and one or two broken-down cars blistering in the sun. The vultures clapped their wings on the roof, and beyond the roofs was the river flashing like ground glass in the middle, and grave with the shadows of the trees at the banks.

A smell of wood smoke and charcoal and rotting oranges came out of yards and the funnel of a launch smoked deserted at the quay—almost the only sign of life. Calcott looked hungrily at Johnson's pocket as they walked up.

It was certain; he had been right. But she could have written. He felt slighted. He craved for her. He craved for her to come there and tell him with her own lips.

Calcott hung about in the courtyard glancing at Harry's window to catch sight of him reading the miraculous English letters. But all he saw was Harry putting them on a table and picking up the two whose envelopes were in Mrs Wright's handwriting and staring at them. Calcott could not bear this and at last came to the window.

'What's the news?' he asked.

'Oh, nothing,' Harry said.

Calcott wetted his lips with desire to see news from England.

'When Blakeney was here'—Calcott referred to an Englishman who had been on one of the plantations years before—'we used to read each other's letters. He had a girl in England and she used to write to him every week. None of these dago girls can write but, God, that girl! I got to know hers off by heart. She was in a restaurant.'

Calcott, always garrulous, told the story at length, his eyes longingly gazing at the unopened envelopes. He laughed.

'The only girl who ever wrote me from the old country was a blackmailing whore who said I'd given her twins. Can you beat it? Me! Twins. Threatened to bring 'em out here. Bloody fool I'd have looked. Have another guess, I said.' Calcott glared and then shrugged his shoulders when he saw this got no response from Harry. With one last hungry glance he went away.

'Class,' Calcott muttered.

The idea of being in a case so closely similar to Calcott's added to the shock of still receiving no letter from Lucy; but again he felt all that was past history, the story of another man. Harry left the house. He went along to the outskirts of the town and sat on a terrace of the land, gazing at the islands which lay flashing like the anchored craft in the summer at Wright's estuary. He did not return for his midday meal.

When at last he came back, the family crowded to his window and Calcott pushed among them.

'Don't want anything,' said Johnson brusquely and shut the window.

But later he was sorry he had been rude and went to apologise to the women. Calcott, ever-present, came out and said, 'Get him something to eat.' They scuttled to it as if a whip had been cracked.

'I've got a message from your pals,' said Calcott. 'They're not coming today. Canoe's rotten. They're up on one of the islands. A chap down at the water-front told me. Last night did them in.'

'When are they coming?'

'Ask me another,' said Calcott. 'Tomorrow. Or the day after. I told Wright the canoe had been lying up in the sun all the summer.'

'Are they all right?'

'Sure, they're all right. It was just the storm. They were on one of the islands.'

Calcott added:

'I told you something had happened, didn't I? When that parrot died.'

Then Harry looked at the letters and to him it seemed that for one day more, or two, he had been reprieved. He went out again to look at the canoe. He thought that if he could get someone to help him with it to the river he would load it up with stores and test it to see how it would behave under a full load. He argued that he must fill his time profitably until Wright and Phillips returned.

Circumstance helped because Silva appeared again.

'Don't you do any work?' Harry asked.

'No,' said Silva. He smiled. He had the curiously detached smile of a native child, full and yet seemingly vacant. He agreed eagerly to get men to carry the boat and the stores to the water-front.

It was a slow job; the afternoon had passed its greatest heat, longer shadows were on the river. One could feel the heat like a tongue on the skin but a little quick flicker of coming coolness began to vary its slow and heavy strokes. Harry took his gun. He might, he said, get a shot at something before sundown. He did not speak much to Silva or the men because his voice trembled; one of the effects of his sickness was to make his voice seem too strong for his body. Or perhaps it was that his thoughts and his body had become disparate: there was no returning strength in his mind but a kind of frozen chaos and his body went on independently of it.

There was no such thing as a private departure from the town. There was always a nondescript crowd at the quay, amiable, grinning, tobacco-chewing people, chiefly men; some half-naked. They came out of the shadows of buildings to watch. Since the rubber industry had gone the shadows of the country seemed to exude this wretched, abandoned and derelict population which moved with the curiosity of a colony of monkeys to the bars of a cage.

Many were laughing maliciously and some of the boatmen got into their boats. Harry saw he was going to have an escort. 'He'll drown'—they wanted to be in at the death. One or two seemed to be wondering whether they might come into possession of the gun. Silva kept off the importunate ones. He was the most voluble and important man on the quay.

There were police on the quay. They stood, two or three of them, smoking dejectedly and jostled by the crowd and patiently rebuking them. When it was clear that Harry was ready to be off the eyes of the police woke from stupor. One of them, a lean bent twig of a man trussed like a fowl in the straps of his accoutrement, came forward and said:

'Friend!'

Harry paid no attention.

'Friend!' repeated the policeman, looking down into the boat.

Harry looked up.

'And the gun, friend?' insinuated the policeman.

'What about it?' Harry asked.

'It's forbidden to take the gun.'

'What's this?' said Silva. 'Who says it's forbidden to take the gun?'

The policeman turned round and studied Silva. Suddenly a furious argument sprang up. Where was the Englishman's permit for the gun? Ah no! Ah yes! Where was it? All the world, the policeman said, must have a permit.

A situation of this kind develops like the problem of the trinity. Schools of thought rise up, casuistries are insinuated, impregnable dogmas appear, a vast theological impasse is created diverting everyone into schisms and heresies. The English intelligence does not flower among these conceits and speculations; it takes the brusque, simple vulgar view of the case. Johnson said, 'How much does the fellow want?'

But the occasion was made for Silva. It was a time when the artist awoke in his blood. He broke with the rapidity of a dull tropical stem into a foliage of argument, marvellous and intricate. His words were like insinuations of perfume which first made the policeman's nostrils twitch uncertainly, then moved his lips and entranced the eyes of his strapped-in face. And Silva, as an employee, though humble, at the Power Station, had almost an esoteric prestige. By the end of his speech, Silva had converted Johnson into a beguiling fiction. He was a Director of the Electric Company. He had a fleet of canvas boats which were to be used by the government. Wherever there is a gun in South America there is lyricism and Silva had made of Johnson a lyrical, mythical figure worthy at the end to stand by that other masterpiece of his—the Hamlet of the seances. Silva himself at the end of his speech turned dazzled eyes upon his model and even when Johnson said, 'Well, what about it? Ten enough for the swine?' the ecstasy of creation was not extinguished. The policeman was dazed. He turned round on the crowd and told them not to push him into the water. He gazed upon Johnson with wonder and softened, not indeed into smiles, but into a man whose importance was blissful and passive to him, almost sighing through his slightly parted lips as if he were being tickled by fingers divine in their discretion.

'Nothing,' said Silva. 'I said I was going to the island with you. I know him. I got his son a job.'

'Do you want to come?' said Harry.

At that moment it was what Silva most wanted in the world.

A dozen other boats shot after them as they paddled out.

The negro sergeant moved away into the crowd like a man in a dream. Silva sat amidships smiling like a small royalty at the courtly glitter of the attendant river. A race was on. The other boats came shouting and singing across Johnson's bows, drifting near with criminal intent, splashing ahead to be first at the sinking place. Gaiety was on the water. But Johnson's canoe was as light as a leaf, Silva gave it a necessary steadiness and with stern low, bows pointing like a bird's throat at the sky and with the breeze on their backs, they paddled on.

And now with the green river wide before him, its far bank nearly a mile distant, the wind lifting the bows like a curled feather, they slithered over the water. There was not a cloud in the sky. The spray spat. He felt the lick of the current on his wrists, the travelling power of the stream. Islands of grass drifted down, slopping and suddy in their trailing wakes. Tree trunks rocked, gulls and plover rode on them and from the islands small birds enamelled in blue and vermillion went by like flecks of crying fire, and herons stood on the shore.

Johnson went on. There must be one more mile, one more depth of healing to be touched.

He knew it was a dangerous boat and would become more dangerous as it became worn and older. It would be dangerous in squalls. He loved the boat for being dangerous. He would love it more as its danger increased. He considered its trembling sides now and its stays of new wood with criticism and content. If he was testing his boat he was also testing himself, gambling, rediscovering his luck.

CHAPTER EIGHT

THE sound of English voices got Calcott out of his hammock and he hurried out of the room. Wright and Phillips, their faces reddened by the sun and their eyes sparkling and dark with the pleasure of their journey, came towards him.

'Seen young Johnson?' asked Calcott.

'Why, no,' said Wright, 'did he go to meet us?'

'That's it,' said Calcott. 'That's just it. Did he? He's not here. He's gone.'

'Where has he gone?' they asked.

'Yes,' said Calcott, 'where has he gone? And Silva's gone too. They went out yesterday in the afternoon and we haven't seen or heard of him since. It was after he got his mail.'

'The devil he did,' said Charles Wright and went to his table where the letters were. He put on his glasses and opened them while Calcott poured out all he knew about Harry's flight. Calcott was afraid he would be made responsible. People come up on expeditions, get lost or in some sort of trouble and the next thing you heard was, 'Well, didn't *you* go up and get them out of it?' As if he was a bloody park-keeper. Calcott gaped at Wright's calm.

'Any news?' Calcott said, looking at the letters.

'I beg your pardon.' Wright looked up in his friendly way, but the shadow or reproof was unmistakeable and Calcott, murmuring 'The boy doesn't know what he's doing,' left the room.

After they had been alone some time Wright put down his letters and said, in a dead voice:

'Harry sounds quite recovered, doesn't he?'

'He got bored,' Phillips said.

'Yes, if you want him to stay in bed you've got to chain him.'

'I think perhaps,' said Wright, 'he ought to have asked the chief? Don't you?'

This was the only evidence that Wright was concerned about

Johnson's disappearance. Their calm, and their entire preoccupation with starting on the expedition straight away, silenced Calcott. He gazed restlessly and miserably at their preparations. They would go off like this, like people going off for a holiday and then he would be alone once more.

'I don't envy you,' he growled.

But he envied them from the bottom of his heart.

That night passed and no news came of Johnson.

But as the morning and the afternoon passed and all that remained for Wright and Phillips to do was to put their stores on the boat, the two men became restless and anxious. They could not sit in the room together; first one went out of the house and then the other followed him like a dog. Wright went down to the quay. They found out that Johnson had not gone to meet them. They went out to the islands to see if he had been there. They were afraid, deeply afraid when they heard about his canoe, that a squall had churned up the river and sunk him. They dared not speak their fear to each other. They joked.

'Tit for tat,' said Phillips. 'He's paying us out for leaving him.'

'Third day of leave overstayed,' said Wright on the third morning. That day a man who had been coming down the river during the last few days, a small planter, arrived with news. He had passed close to Johnson's camp in the evening and seen his fire. Two other men confirmed this. They had seen Johnson two days before and still travelling upstream. He had passed out of sight. Later, echoing over the waters and down the forest wall, they had heard a shot.

Wright stuck out his bearded chin.

'Damn him,' he said. He was white with anger. Yet in their different ways, Wright and Phillips were admiring Johnson for giving them the slip and for making this mystery.

'Wait till we catch that young bastard!' Wright exclaimed, looking into the milky foam of the early mists on the water. He was on his mettle.

But Phillips was not angry, nor had he any cause to be. He was apprehensive. His vanity was wounded that Johnson should have fled from him. And when he laughed it was with less irony and more fully than Wright first of all. The anarchy of Johnson's behaviour was what pleased him.

'Let him go,' he thought. 'He has a right to go if he wants. We all have a right to do what we like. Let's break all the rules.'

This was a heroic rationalisation of his own personal fears of the journey itself. He had imagined in England a long journey on foot and on horseback. It was to be by water; and he disliked and feared the green snake smoothness of water and the translucent depths of the river where its beds became white sand, and he himself seemed to be mounted on a frail and perilous glass which might, at any moment, break beneath his weight and plunge him choking to the bottom.

'I'm against all this water,' he said to Wright. Wright was in no mood for Phillips. Their boat was a heavy one and they had a crew of four men at the short oars. There was a thatched deckhouse in the stern and the men rowed forward. The stores were amidships. They had a lighter boat in tow.

'I'd give five pounds to see that lot sink,' Calcott said, watching the long heavy boat when they left him. When he could see them no longer from the quay he went up to higher ground and watched them. They were passing like a black insect between two islands. They had come, they had broken away, he did not know where he was with his peculiar countrymen. He went back to his house and stared into the rooms where they had been. A bottle of whisky, scarcely touched, was the only tangible thing he could recall in connection with them. He went off and found it and took it down to the power station.

A week passed. The two men rolled themselves in blankets on the sands where the camp was set each night and lay listening to the chatter of the men. It was strange to camp without Harry.

By day they paused at the Indian villages, filthy places of greasy smoking fires and dilapidated huts, to get news of him. Sometimes they heard of him. He seemed very near but they realised he had got a good lead. They drank their coffee. They cooked their black beans, they mixed in the tasteless farinha meal, they lived off their guns. Dark green forest walls gave place to wilderness of scrub and then forest returned again. Rain

98

soaked them. Sometimes the boats of Indians passed them, whole families in limp migration on the stream or fishermen shooting their arrows into the water. The Englishmen's nerves were on edge; it was infuriating to think that Johnson was ahead of them and had, so to speak, sent these sights down second-hand.

On one sandbank by chance they came upon one of his fires. A tobacco tin was there but the fire was cold and the birds had cleaned up all remnants.

'He might have left a note,' Phillips said.

'Less wit!' Wright replied shortly.

They had exhausted the topic of Johnson. They could not imagine the reason for his behaviour; but preoccupation with him underlay all the pleasure of this life of camp and the casual adventures of each idle, sun-filled, monotonous day. Bitterness came into Wright's baffled talk about him. Phillips could not but sympathise with Wright but had the tact not to show his sympathy. They were in an absurd position. Wright was leading the expedition and yet he was following. Or there were two expeditions? Or—what was it? But when they talked of the next day's plans or of the ultimate object of their journey—the journey on foot—and whether they would get any of the listless and greedy Indians to guide them from one river-head to the next when they came into really untouched country, Wright spoke as if nothing had been altered; as if what they had discussed in his house, the adventure of his middle age, his defiance of the thought that he was no longer a young man, remained inflexible.

And the humiliation and anger had not only spiritual causes.

'Never known him rat like this. Must be something,' Wright kept saying. 'There's an explanation somewhere.' 'Cherchez la femme,' said Phillips. He did not mean to say this. He was one of those young men who have a lot of tags, quips and quotations on their restless tongues. Charles Wright was getting tired of them. But he was sufficiently experienced a traveller to know how easily one gets on edge and finds petty faults in one's companions.

'Is there a woman?' asked Wright.

'Lucy was a bit gone on Harry,' Phillips said. 'I thought he was. But I didn't mean that.'

'I didn't know,' said Wright quietly.

'I thought,' said Phillips with a little confusion, 'you must have observed.'

Wright said he had not and they talked no more about Harry. No one would have been more delighted than he if Lucy had talked of marrying Harry; no one more surprised if such a thing happened. But he was a little put out that he had not noticed this warmth. Not having noticed it made him feel old. How secretive the young were! He had never worried about his age for he knew his heart was young, but this gossip joined Johnson's flight and Phillips' furtiveness to mark him off from the young. Time has planted, not a gulf between the young and the middle-aged, but a wilderness. He would have felt equal to a desertion of men of his own age but betrayal by the young is another matter. That seemed to indicate boredom.

There was another mystery in this for Charles Wright. Harry had got Silva the Portuguese with him. This man Wright remembered in his earlier expedition. He had been taken on by the gold-prospectors. Why had Harry taken him? Was Harry purposely stealing a march on them under Silva's guidance? Wright was tortured by this; for he was too noble and simple a man to suspect readily and he was far more unhappy because of doubts and suspicions which attacked him, than because of Harry's desertion.

Their boat was heavy. Johnson's was light. There were four men at the extraordinary oars of their boats—which Phillips called the sewage scoops—and the two Englishmen had tried to manage them. It took time to get any skill and the work was monotonous. Their hands were cut and blistered by the rough poles. Again they lost time by the scrupulousness of the enquiries, for they might easily have passed Johnson on the river. The greater part of one morning was lost because the crew had gone up to a village the night before and found drink and women. They came back sullen, exhausted and quarrelsome. They made excuses to stop rowing in the fierce sun. Wright's usually persuasive patience left him and he made the mistake of showing his anger. 'And do stop humming and staring like a fool,' he cried to Phillips. Phillips apologised. When he had done this he realised at once that Wright would have preferred him not to apologise.

But this did not reconcile the four Brazilians. That day had to be lost and a better beginning was made the next day, when the Brazilians had stopped talking about the women: and they smiled again when Wright promised them beer the day they caught up Johnson.

But on the ninth day they heard news which brought the pursuit to a crisis.

It was in the evening. They had encamped on a sandbank two miles long and the crew were in a good humour because they had found turtle-tracks in the sand and had traced them to the place where the eggs were buried. They had dug out forty-five eggs. In the middle of preparations for the meal, Phillips saw a tall man dressed only in a shirt walking across the sand towards them. He was dark against the setting sun and for a moment they thought he was Johnson. He called to Wright and they stood waiting for the stranger. He was a tall European with lobster-coloured skin and a short ragged fair beard.

'Good day,' he said in Portuguese, standing stiffly, grotesquely formal in his fluttering shirt. 'My name is Schauer. I'm a Czech,' he said. 'I see your camp smoke.'

'Do come in,' said Phillips. 'You've come at the right moment. We'll lay a place for you.'

The enormous Czech sat down in the sand. He explained that he had come down that afternoon.

'From the rapids?' asked Wright.

'What is that?' asked the Czech politely. His English was poor.

'You know—the rapids, the . . .' Phillips made absurd gestures.

'Oh no,' said the Czech. 'From the little river.'

'Have you seen a man in a canvas canoe?' they asked at once.

It appeared he had seen Johnson two days earlier but on 'the little river.'

'He's gone the wrong way,' said Wright dumbfounded.

Now, indeed, Johnson was leading the expedition.

All the evening they feasted the Czech on turtles' eggs with sugar, and they sat picking ticks out of their skin; but the question 'Why *the little river*?' did not leave their minds. Johnson had not mentioned his pursuers to the Czech. They spent the night talking together in muddled English about the country ahead and then had parted. He had plenty of food, the Czech said.

'Yet he knew,' Wright said to Phillips, 'that we are not going up that river.'

'He likes travelling alone,' Phillips said. 'When we were coming up to you, do you remember I told you he said it would save time if we cut across country? I said to him, "You can't go through this forest very far. There are no roads." But he said, "The Indians get through. One could go from tribe to tribe. They're not hostile. Even the so-called hostile ones are all right if you manage them. Anyway," he said, "my idea would be to keep clear of them." Of course, he didn't do it but I thought to myself, "If I weren't here he might." I supose the idea just occurs to him. When someone says something is impossible he wants to do it. He likes difficulty. Directly the difficulty becomes simple he is happy for a moment because it proves what he always said, then he starts adding a new difficulty.'

They had been talking in the dry cool darkness and the sad, piping cries of the river birds and the cold lap of the river on the shore came between their words. The silence of other nights had gone because now the sounds of the rapids miles distant came to them along the river, a moan like the organ sound of a steady wind in the forest. There were places where the current had been swift and they had kept to the shore most of the day.

To be alone. But Wright, hearing the prolonged diapason of the rapids, knew it would be difficult for a man to pass alone in a canvas boat when he did not know the river. A man like Silva would desert, and Wright thought it was doubtful if even the converted or pacified Indians of the river villages could resist for long the chance of attacking a man alone who knew nothing of their language except a few odd words.

Johnson's smattering of Guarani would not take him far. When men are alone they begin to talk to themselves and create worlds of habit and fantasy in which they retreat from the real world. It is the beginning of madness. But he knew that the conditions of the country, the conditions of solitude, were ultimately beyond luck and the purely reckless impulse and that the chance of Johnson being killed by accident or getting ill and dying, dying of thirst or poison or exhaustion, were very great. And Wright forgot his own wounded pride, and his anger, in his anxiety to save the young man.

When they set out again and were themselves in sight of the rock islands and the swift water of the rapids, stopping with some Indians who came whining and squeaking like bats to the shore, they heard that Johnson had passed the rapids. He and Silva had managed it.

BOOK THREE

CHAPTER NINE

DESERTION is one of the commonest occurrences among parties of exploration. A man falls sick, his nerve fails, there are quarrels about objective, direction, time or money, there is jealousy of a leader, weakness or tyranny in authority; again in isolation petty idiosyncratic differences become magnified into intolerable crimes. Those who take the heroic view of men who go into new country are shocked when on closer view they see their gods behaving like a nunnery; sides are taken; there is talk of betrayer and betrayed.

Every possible reason has been brought forward to explain the desertion of Wright by Harry Johnson. Two of the most fantastic, so fantastic that they could easily be disproved, happened to come near the truth. They came from Silva, an artist in such matters but with the artist's irresponsibility, his weakness for getting his facts slightly wrong. Silva said that Johnson had two grudges against Wright: one that Wright had not told him that the real object of the expedition was to discover the gold which Wright knew of from his earlier expedition. Wright, said Silva, had deceived his friends. He had been in the country weeks before either of them and had been at a mineralogists' conference at Caracas. The proposal had been made to him there and he and the mining company were going to get the benefit of it.

This was untrue, but there is no doubt that while they were together Silva suggested such a story to Harry Johnson and that they discussed the probity of Wright.

The second alleged cause of quarrel between Wright and Johnson was in Silva's best melodramatic manner; it was that Johnson was in love with Charles Wright's wife and that she had just had a child by him. Before his flight he had had letters from her. Calcott and Phillips, too, both confirmed that Harry had been exercised about letters. Neither of these stories has appeared in any book on the Wright expedition, but a sensational American paper got hold of the latter story and published it. The article was never reproduced in England; but when grave rumours about Johnson and Wright got about, they made, with this earlier episode, a very plausible story. Phillips' diaries say nothing of these things. He records simply: 'Harry gone.' 'No news of Harry.' And so on. He speculates and then drifts into his own sensations. He is accurate but unrevealing. Phillips was not, however, suppressing what he knew; like so many diarists, he was overwhelmed by the big event. The more he felt the less he wrote. And there is another reason: he was overwhelmed, and for the first time in his life, by something that had occurred, not to himself, but to somebody else. By a paradox, it is not from Phillips' accuracies that a reader would learn what happened at this time, but from Silva's fantasies. The creator of Hamlet had once more merely fantasticated the truth.

For Harry Johnson did not come back that evening. He intended to return and yet he went on and on up the river knowing that night would come suddenly and he would be too far to go back. He camped for the night on a sandbank and took draughts of freedom from the night. Silva was no difficulty; rather he justified the act. The creator of a new life requires a witness. The nonchalance and irresponsibility of Silva was a quality which gave a lightness to the undertaking.

And he was the most delicate tempter. In the morning Johnson's first thought was that Wright and Phillips would return; Silva unknowingly (or perhaps knowingly?) diverted his thoughts. He began to boast of his strength

'I can swim far,' he said. To show this he took off his clothes and went into the river. He was pot-bellied, dark-skinned and very hairy. There was a splash and away he swam.

'You see,' he said. He crawled out of the water. He pointed to the muscles on his arms and his thighs and then performed small gymnastic acts on the sands. When he had done this he said like a child:

'I come with you?'

The startling thing about Silva was the way he apparently read that part of Johnson which was pressing him not to go back.

'And your job?' said Johnson. 'What will Mr Calcott say?'

'Mr Calcott! He is a fool. He is old and mad. I cannot go back to work under that man.'

There was your Silva! He did not even consider the question of obligations. He stood there, sweating and nonchalant with his two days' growth of beard, his eyes ringed with sleepiness, and his smooth hair roughened by the wind.

'I go with you,' said Silva simply.

'Where?' asked Johnson with uneasy amusement.

'Anywhere. The way your father went,' Silva said in his perfunctory way. Johnson involuntarily turned and looked at the river as though expecting to see his father pass by in the sun of seventeen years ago.

A curious vividness was given to everything Silva said by the imperfection of his English. The faults of accentuation, or a word misplaced, strengthened the words with their streak of the bizarre, gave a suggestion of ulterior penetration and meaning. The birds of the river whistled and called like boys. Their hardness and fullness and boldness, unlike the softer chatter of English birds, was a language; Silva's speech had the same arresting, exotic quality. It was all the more arresting when Silva spoke about the dead missionary passing along the river.

His voice was like the voice of the free, wild country. It tempted not the mind but those instincts that quiver at new sounds.

Johnson looked at the glittering sweep of the river leading into deeper and deeper hazes and recesses of heat that would be like ever fiercer ovens of guilt as he passed back through them to Wright. And then he turned to Silva and the air was younger and fresher and there was no guilt. Because Silva did not know him, because not even instinctively could he guess, because he was foreign. Johnson had grown to crave to be alone, not to feel the minds of people cast intangibly upon him with the spidery lightness of a net. The barrier of a race different from his own lay beneficently between Silva and himself.

'Going to shoot something,' muttered Johnson and trudged

across the sand to the trees two or three hundred yards away. The morning was passing. Silva dozing on the sand was awakened by shots and the rising of a hundred birds out of the trees and the flight of white birds from the sand crying over the river. Johnson came back with his bag. He flung the quails into the boat and said:

'Dinner. We'll push off.'

He was heavy-browed with thought.

Harry Johnson was a reasonable man. But, above all, a reasonable man and very balanced. Judicious, carefully weighing the pros and the cons. At thirteen years of age, in common with thousands of other young Englishmen, he had heard his first Public School master tell him that one played for the side and not for oneself, that one was loyal to the House, the School and the King. But that it was not done to mention these convictions. Like every intensely individual creature he silently revolted from the code. Admirable for the average man, it seemed at variance with most of his instincts. He respected it only when it seemed to be the common sense; and each occasion had to be judged for itself.

But the thing he had most feared in loving a woman had come upon him: it had overthrown his reason. He had found himself in a world of naked instincts, and the appetite of the instincts increased with eating. He had discovered how strong his desire for self-torture was. After he had been Lucy's lover for a time and she was congratulating herself on rescuing a man from the marked self-loving fantasies of the solitary, he found that his torture fantasies had not disappeared—as she imagined they had—but that now he wished to make them real. Lucy and he had quarrelled.

After he had become her lover this was the perpetual subject of their quarrels. They started when one day in a London street he saw a man roped to a chair.

'Wait,' said Harry. 'You watch.' They saw the man step easily out of the knots that bound him. Lucy marvelled. She and Harry were in love and, wandering together in the streets, they were always seeing things to marvel at. Every sight was tinged with this miraculous quality. But Harry said, 'That is not as difficult as it seems. I'll show you when we get back.' When they got back to her flat he took a rope from his suitcase. It was the same old sheet which he had once exchanged in Wright's boat.

'Here,' he said. 'I'll show you.'

He sat on a chair and held out the rope.

'Come on,' he said.

Lucy was a little disturbed and incredulous and, after some argument, did as he told her. It was only after a few minutes, after seeing the look in Harry's eyes, that she suspected this imitation of the trick they had seen in the street was a pretext.

He was ashamed when he saw her horror; that is, his mind was ashamed, but instinct is non-moral and unrepentant. And it was the more powerful for being brought to the surface. They quarrelled and their quarrel ended, under Lucy's guidance, in confession. Because he was abnormal, he said, he had taken to the life he lived. He went to live alone; his journeys were a mortification of the flesh, a self-torture in flight to the ineffable sensations of the spirit.

The guilt he had felt because of Lucy was really another guilt in disguise; it was also the fruit of his dissatisfaction with her: 'She is having a child; that is my punishment (and her punishment) because she is not a satisfying woman to me.'

And now on the brink of his flight from Charles Wright, because of Lucy, the same motive of self-punishment was at work. It was coupled with the instinct which makes a man punish one before whom he feels guilty. You wrong a man once; and you wrong him again as a revenge for making you feel guilty. In making you feel guilty, he wrongs you. You are justified in wronging him in return; not morally justified. Morality does not come into the matter; it is inevitable.

So one guilt wipes out another. A murder wipes out a theft. To abandon Wright was against all his tradition and his reason; but in guilt, if it is strong enough and when it is aided by deeper guilt in one's nature and the aspiration to freedom which springs perhaps from the very depths of guilt, tradition and reason are nothing.

Silva and he pushed off the boat and paddled upstream. And here again Silva appeared to be the ideal unwitting accomplice; he had fertilised the seed of flight. Now he was caring for it. He talked incessantly. They went on under the protection of his chatter. A look of deep content was on his face.

'You'll have to work,' Johnson said.

Silva took a paddle.

Johnson looked behind them often.

'They'll pick us up,' he said, 'when they get back.' Still he had not admitted his desertion.

Sometimes he smiled sardonically at the thought of pursuit. He was still elated on the water. That night at their encampment when the darkness came, Harry lay in the glow of the fire and Silva was astonished to hear him start singing. *Onward Christian Soldiers*, *Tipperary*, and *Less than the Dust* were among the songs. Silva, after tactful listening, sang two or three Brazilian songs, but Johnson interrupted him and started singing again. His voice sounded small and strained in the wide river air and he broke off if a fish jumped or a tree cracked or a night bird called.

'This bloody country's too big. It's got too much of everything,' Johnson said suddenly. This was the only sign that he had even noticed the country.

Silva understood that Johnson was singing with sadness and anger and did not follow him when he got up and walked away for two hours to be alone. When he came back he said:

'You say, Silva, you know where my father went.'

'I know where he left the river,' said Silva.

He pointed vaguely in the darkness.

'He must have wanted a job,' Harry said.

He spoke abruptly and hastily to Silva, treating him with that laconic contempt which conceals the gratitude one feels to an accomplice.

They got on fast the next day and, getting ashore in the evening, carried the canoe on their heads past the rapids. Silva said little and sweated under his load. But he did not complain.

'We stop here,' he said, concealing his weakness.

'No,' said Johnson. 'Got to get on.'

He had been going to stop, but he had to be the taskmaster to Silva. They ate a cold greasy meal that night in the darkness and in the middle of the night there was rain, heavy, driving rods of rain. They were soaked and shivering in the morning. Silva was miserable. He stood about pouting at the river. But coffee and food revived him and the sun dried them. He began to boast about 'a little drop of rain' hurting no one. He spoke always as though at any moment he might walk away and leave Johnson out of absentmindedness.

Johnson liked this and Silva like the abrupt unrevealing stubbornness of Johnson. There was a pretence between them that they were just going on a little further to see what was round the next bend, but they both knew otherwise; that they were now completely governed by a decision to go on alone. Silva, to show this, fished and cooked.

A deep friendship was forming between the two men. And they did not discuss their situation. They knew what they were doing. Silva knew that Johnson did not want to see his companions and this seemed unremarkable to him. One must always do what one is compelled to do. It had occurred to Silva often not to wish to see his friends again.

A smaller river joined the longer one in the afternoon. They had kept out from the banks which here broke up into creeks where the alligators belched and grunted, for they did not want one of the beasts under the fragile boat. The forest overhung the banks and trailed its curtains of liana over the water, and as the two men fought over the boil of the current into the small river the whole climate of the country seemed to change. The air was hotter, and they passed into a rich wilder scene. On the banks fallen trees damned up rotting vegetation in the current, the water was as dark as tea, and the alligators lay with their yellow mouths propped wide open, fixed in a hot-house trance. The sky was less wide here and the breezes were less fresh and more fitful and the river was stamped with a humming monotony. The banks being closer, Harry and Silva were more aware of the jungle. It had the appearance of a bedraggled palisade, smashed in at times as if a lorry had crashed into it, or suddenly it thinned away to low dirty scrub the colour of verdigris.

Silva did not know that in going up this river they were leaving Wright's route. Silva knew only that this must be the route Johnson's father had taken.

But Silva saw a change in Johnson that evening. A new alertness appeared in his heavy features, the smiling laziness went out of his voice. His paddling was less fitful, and all the afternoon his arm dipped and splashed. Silva was tired and stopped but Johnson went on breaking the smooth tea-coloured water with endless energy. They had to go on far up the winding river, which split into narrow channels where the trees met overhead making arcades of quivering river light. The calls of the birds echoed here in the halls of the trees and the alligators sank by the banks. They had at times to beat them off with their paddles. It was not simple to choose the right channel, for some became suddenly shallow and were dammed up by fallen trees and the canoe had to be turned while they retraced their path and tried again. In the shallow water they saw long and vivid fish dart like hands under the boat. Johnson put Silva on to fish. They camped that night in a break in the trees for there was no clear clean stretch of sand, and their skin was pitted with the bites of flies and pinheads of blood appeared on their arms. Silva talked about snakes and scorpions and slept in a hammock slung between two young trees but Johnson slept on the ground. They were tortured that night by the mosquitoes.

Their boat went on through the rustling water and the eddies in the paddle-pits were left like fading heel-marks on the surface of the river behind them. Johnson said as if speaking to himself:

'He went up this river to the rapids,' he said, 'and above them there is a new river on the right bank. You say he went up there but his canoe was found hidden under the bank in the way the Indians hide theirs? Some Indians had come down there because they talked of him afterwards and one of them went with him through the jungle making across the horseshoe to the river beyond it. He must have been cutting across country. It rises to the high land where no one has been. His fires were seen. There are three tribes of Indians, one after another, and they won't go beyond their own territory because they are afraid of being murdered. So he had to go on alone.'

Silva said, 'You are going there?'

'I don't know,' said Johnson.

He did not know. He was only growing into his freedom. There is an insect in the instincts whose feelers push out and touch the unknown with infinitely delicate strokes, cautiously. All Johnson's training was against such an unplanned journey. He knew time must be calculated, stores rationed, ammunition saved. He must know about water. He was in that bemused condition of curiosity and dream which is the prelude to decision and action; the year is like this at the beginning of spring.

He had little interest in the country itself. The forest, the scrub, the marshes, the disorderly, untouched wealth of it did not arouse a response in him—except, as he had said, there was too much of it.

Birds were of two kinds: those he could shoot, those he could not. Of the dangerous creatures, the alligators and the blood-sucking fish which gutted their victims or stung them with poisons, the rattlesnakes, the rare jaguar, he had no great fear. His imagination was not upon these things. Each would be dealt with as it appeared. He did not think for one moment that he would die or could be injured. He rather prided himself in not taking simple precautions.

In his heart he disliked the country; it was too green, too profuse, too monotonous. The Indians were like so many fleas in a rug. He often thought of his last year's journey in Greenland with longing. He longed for mountains and snow. He had listened with deep emotion when Wright had talked of the Himalayas and their whiteness. That whiteness could enslave him; not this greeness. It oppressed him to think of his father throwing himself away on this rampant drainage-system of forest and turgid rivers.

No, this land enclosed him in himself. He was not travelling as he had travelled in Greenland; he was travelling here in himself, paddling down the streams of his own life and nature, enclosed in the jungle of his own unknown or half-known thoughts and impulses. But present with him all the day, written on the walls of the trees in all their variegated detail, was his own life ramified, overgrown, dense and intricate and mysterious in its full tones, half tones and shades of consciousness. The forest itself was like the confusing, shapeless product of a torpid and bemused introspection.

Lucy had gone. Wright was distant, an outdistanced pursuer of conscience. Phillips was an embarrassment and never important. Silva was a man, and yet more like a child, a creature who had the disturbing faculty of speaking aloud Johnson's own buried thoughts. But Silva was not more than an intruding voice, an indicator. Slowly as they went past the trees, Johnson began to exchange the presence of Wright, Phillips and Lucy for the presence of his father.

Especially in the glazed heat of the afternoons when the shadows began to slip like caps and gloves upon the trees of one bank and there was no sound but the sound of the paddles, Johnson would be arrested by the silence of the trees, by the sensation of being watched. Sometimes he might imagine the tree itself was looking at him, that its trunk and spread of branches were in some fantastic, struck-rigid way, human; or that people were concealed in it, or animals, their eyes as fine and narrow as needles glinting with note taken of him. There was the sensation of casually remembering some incident in one's life and of writing it upon the faces of the trees, spreading one's life on them, and then on finding one's memories breathed out thickly, heavy with brooding and weighted with significance, upon oneself. The trees would hold every thought one had had and would keep them there for ever, so that looking back one would dread to return that way, here to meet this shame or that voluptuousness or hatred. And the trees ahead and the trees sliding smoothly past, moved slowly like heads chin-down upon the water, brooding, so that one expected a deep voice to come out of them.

But the sound that came out of them was not the deep voice in his memory, but the sudden, motor-horn bark of the toucan in the tops of the trees, and pure high outlandish calls from the river birds, startling, fresh and mocking in their irrelevance.

The voice he half expected was the voice of his father. An inward awe of this region where his father had passed was growing in Johnson's heart. As they paddled to the shore in the dusk, forced to camp in the mosquito-infested trees, a feeling had possessed him that he was near his father; and as his foot touched the ground, that there was a danger in standing upon the land where he had stood.

'The Indian was probably with him still,' Harry said to Silva.

Since Silva had mentioned the missionary Johnson spoke of him and they argued about the seance in Calcott's house.

'He had no right to do that,' said Johnson. 'I didn't actually mind, but some people would have minded—people who believe that stuff. You don't believe it?'

'Oh no,' said Silva. 'I think he is dead.'

'Of course he died.'

Johnson said after a pause:

'When you are alone like this you get a feeling that he is still alive. It is the silence.'

'It is possible,' said Silva.

'No,' said Johnson. 'It is because no one saw him die. If his death were not mysterious I shouldn't feel this.'

The river was like pale marble and the trees stood straight and black plumed by it. Johnson dreamed that night that his father came down the river in a canoe and landed at the camp. He had only one arm and his eyes were accusing. Phillips was with him and there was a woman sitting in the canoe.

In the morning they lit their fire. They had woken stiff and shivering. Silva took out the boat and fished. Johnson examined the diminishing stores. They were only enough, after they had bought blocks of rapadura and farinha in the last village on the big river, for another ten days. 'My father must have lived off the country.'

Weakness overcame them on this day. Johnson felt giddy and sick. He went away in the trees and vomited. He suddenly had a horror of their camp and they broke it up and went on. The land swam. His body ached. They lay half the day in the shade of the scrub and Silva watched him.

'I'm going on. In a minute,' said Johnson at intervals.

He was suddenly shocked by his desertion of Wright and Phillips. He hated the sight of the glum, tired-eyed Silva. He was horrified by the morbidity of his thoughts about his father on the previous day. He lay thinking about Lucy again. Exorcised she had rushed back again into his mind, loved, desired, longed for. He tried to remember her voice. He remembered her skin. He remembered how they took off their clothes. He dreamed of slipping back in the boat, eluding Wright and going straight back to England. An avid heat glowed in his mind. He sat up; he saw the scene, hard, brilliant and new, no voluptuous dream of the solitary. He looked around for Silva.

'Silva!' he called.

Silva came up. Harry told Silva about Lucy that night, not very clearly. It was not like speaking to a real person because Silva was a foreigner whose accent made him seem unreal. And Silva, the hero of so many adventures with women, scarcely listened. He was burning to tell some of his own stories. A subject on which he could at last be eloquent with Johnson had come up. It was due to this impatience of his that he got the impression that Wright's wife and not his stepdaughter was in question.

CHAPTER TEN

THERE was a late moon and the raiment of water, dividing the trees, made a scene of metals. Animals cried out in the forest through the night. There had been laughter about hostile Indians during the day, but now Wright and Phillips watched the yellow flame of a fire, no more than a scratch of yellow, on the distant bend of the river.

They were, they calculated, only a day and a half's journey from him. They slept uneasily under the white arc of the moon, and, after the usual sullen grumbling from the men, drank their coffee and started soon after sunrise. They sat advancing into the dazzling sun.

There was a monotony in the brief but overwhelming youth of these tropical dawns, when the land lay without shape like a divine breath upon the air. Phillips leaned eagerly forward and Wright stood keen and grey. His beard seemed to stiffen and his eyes, half closed against the new light, glittered like slits of dew. There was the smell of the trees in the water and the smell of the sleep sweat of the crew. There is to an ageing man nothing more cruel than the everlasting youth of the world, and it was not only the need of travelling fast and overtaking Johnson that made Wright stand up in the morning and impatiently urge on the crew, finally getting to work himself with the paddles; the impulse came also from the exquisite pangs of an envy for Johnson's youth, a desire to reclaim his own. Wright's temper sharpened and the crew grew sullen.

'Can't he see that's not the way to get these fellows to work,' thought Phillips.

Phillips was a man not used to obeying or being obeyed.

As for Phillips, he saw Johnson in his camps and he saw the courage of Johnson. Not in imaginary encounters—there were pictures made for himself by his own fears—but he saw Johnson, still alone, ordinary, unthinking. The essence of the courageous man's life is that nothing happens to him. Phillips took the paddle and felt the boat shoot on to Johnson's courage.

Suddenly Wright stopped paddling and called out.

'There's someone there,' he pointed to the far bank.

Two figures were moving against the trees.

'It's Harry.'

They all stopped and stood up in the drifting boat. Phillips put his hand to his mouth and shouted. The shout fell on the water and the forest. The two figures stood still on the shore. They made no sign.

'The police launch then went ashore,' Phillips began.

Wright said 'Shut up. You're not to say anything like that to him.'

Nearer and nearer the two figures came.

Silva and Johnson were standing there.

They were standing some yards from the water's edge, when Wright landed and walked up to them, two scrubbily bearded men with their clothes dirty and torn, the skin on their faces and their arms reddened and spotted and swollen. Johnson murmured something to Silva and they both grinned. They were looking at Phillips and not at Wright. Johnson said:

'Hullo. You know Silva. He's been fishing.'

'Of course, I remember him,' said Wright genially stretching a hand to Silva.

The crew came ashore and the four pretended to study them. Then Silva came forward and broke the awkward greetings:

'I will make us all some coffee.'

They sat down on the ground and Wright talked of his journey. 'It was my fire you saw last night. There are no Indians,' Johnson said.

'We thought it was Indians cooking you,' said Wright.

'There was a nasty smell in the air,' Phillips said.

'Mosquitoes were our only trouble,' Johnson said. He showed his swollen hands.

They gazed at Johnson and felt a deep affection for him in his comical situation, wondering how he would brazen it out, longing for him to do so. They prepared to laugh loudly at him, to heal the strange breach with laughter. But Johnson, like themselves, gave no hint. There was a set expression on his face of an enclosed man who would not explain. Wright's diplomacy was a diplomacy of suggestion: 'You've proved your theory about the boat,' or 'What do you think about this river? It strikes me as being useful. The wrong river is often the best.'

But all Johnson revealed was that he had no opinion about the river yet, because he had been ill for two days and they would have been much further up but for this.

'Food was getting short,' he said.

He seemed to suggest that but for food they would not have caught him.

But as time passed Wright was beginning to lose interest in the comedy of their situation. Johnson was no longer symbol of youth or courage. The sun was at midday. Wright was a man of forty-nine. He had planned his purpose on the meridian of his life. He said quietly:

'You know we shall have to go back to the big river according to the plan we all agreed upon.'

As he began speaking Phillips got up and called Silva away with him.

Wright continued as they went out of earshot.

'I think your effort was a magnificent one, but we must work as a team and a sideline like this would waste time and we've lost too much as it is. I knew of course you were trying the canoe. Calcott got melodramatic about it but Phillips and I didn't worry. You shifted too! But now we've got to get back. We're a team, Harry.'

Wright waited for Johnson to speak. At last Johnson said:

'I think you'd better send me back.'

'What do you mean?'

'I'm sick.'

'What is it?'

Johnson mumbled and then said:

'I've lost my nerve.'

'Not sleeping?' said Wright.

'I've slept all right,' said Johnson. His heavy shaggy head turned away from Wright. 'It's my nerve.'

'You've done too much,' said Wright. 'You had a start on us but you got away. You must have paddled like hell. You're just done in and want a rest.'

Johnson did not answer.

'We've got the time,' lied Wright. It is, Wright knew, a common delusion of men who spend their lives in exciting action that their nerve is going. It is an involuntary indecency of the spirit which, Wright knew, cannot be helped. The two men glanced at each other. They had known each other for years. They had climbed together, sailed together. They had described their doings, they had argued into the night. They had laughed at each other. Wright had the faculty of putting the young at their ease, of being merely the man who had lived longer, pretending with a skill they did not notice that this was a disadvantage. He listened. Yet now Johnson and Wright looked at each other with incomprehension. Engrossed in action, they knew each other's idiosyncrasies only.

'I've been looking at the maps to see if we could cut across-country,' said Wright, cunningly trading on Johnson's passion for action. 'When you're fit we'll have a look.'

Johnson did not answer. His face was heavy, stubborn and inert. Wright persisted and went into the pros and cons of the journey, knowing it to be impossible. He talked a long time slowly coming to the judgement that it would be better to return to the main river. Johnson lived every moment of the deep humiliation of the return. He felt only the humiliation of being trapped, but how trapped or why trapped he did not know. He watched Silva wandering along the river's edge and thought of the days of freedom and the chains which Wright's quiet, even voice put upon him. He struggled to remember why it was he had run away from the expedition and could not remember. He would only remain defiant and enclosed, with a growing barrier of resentment against Wright in his heart.

'He used to flirt with young women under his wife's nose and make her jealous'—this irrelevant thought came into his head.

'Silva!' called Johnson, 'Silva!'

'Yes?'

'Are both paddles in the boat?'

'Yes.' To show Silva was his man.

This interrupted Wright.

'Extra crew will be useful,' Wright said.

Johnson's resentment grew at this appropriation. He got up and surprised Wright by walking away. He walked out of the camp and then out of sight of it, and at once his gloom lifted. His eyes were alert, his face ready and lifted, his body waking into its extraordinary agility. The mad idea of going on alone, just as he was without food or arms, hovered in his head. Suddenly he stood still and, amazed at himself, broke into tears. They came without the feeling and without warning and without meaning and the present moment seemed to melt from him like wax under a flame. Wright's bearded head appeared and went. His father's face came. He was, for a powerless moment, a child again and shouting at his father angrily, 'The next time you cross the level crossing I hope the train comes along and kills you.' This was a clear memory dislodged from the time when he was six. He had not wept for years and this memory jumped forward with the suddenness of the tears.

The tears were few. It was as if they had confessed; when his lips had been unable to speak they had spoken for him.

An air of embarrassment was in the camp. The idle crew sat under the trees gambling and quarrelling mildly and watching the Englishmen. Wright slept and Phillips tried unavailingly to draw out Johnson. Silva, observing everything, imagined that there was a quarrel about the division of the sale of the gold. Or perhaps about its whereabouts. Silva reckoned that if the expedition split, he would have a half-share with Johnson if he stuck to him. This was in Silva's imagination. In reality he did not believe the expedition was really going in search of gold but his was a mind whose fantasies never rested. He missed—it was the real hardship of the journey—his cigars. Phillips said to himself, 'If Johnson stays, I shall stay.' There was no reason for the expedition if Johnson were out of it. Phillips was depressed by the boredom of the daily camps and of this camp in particular.

He could feel the slow turning of the earth, the irretrievable passage of time in his life. He thought chiefly of traffic and hot streets and restaurants. He traced the services of buses across the stream. While the others slept he propped up a mirror and shaved off several days of beard, admiring himself as he did this and sighing at the sleepers. He went over to Silva and nodded to them all.

'Crisis,' he said. 'Is Johnson mad?'

Silva shrugged his shoulders.

'Oh no,' Silva said.

For Wright now the incident of the chase was closed and Johnson's 'nerves' were already written off. Brusquer in speech now he was inactive, Wright was also bluntly decisive. Phillips had noted the change from his English manner of quiet, gay courtesy. Wright had the fever of his adventure but he was not one of those who believe in leading and commanding. If he had had twenty men he would have gone on his own way, leaving the others to imitate his diligence and his persistence. His orders were no more than sly digs, fragments of mockery.

Before sundown he said to Johnson:

'Let's try your boat and have a shot at something.'

They took their guns and got into the boat.

It was a late afternoon like all the rest, the heat of the sun lessening with every beat, the distant trees softening in tone and hardening in outline. The two paddled with little noise, keeping a look-out for floating trees as they went near the wreckage of the banks. Wright spoke little and Johnson not at all.

Wright discovered one thing: that Johnson had intended going on. With the silver path of light between the deep shadows of the evening trees tranquil before him, broken only by the rising birds, Wright understood Johnson's wish. He said:

'I don't blame you for wanting to leave us. And I'm sorry to have to claim you back.'

He spoke frankly for the first time; he felt there was no fear of injuring Johnson by the words.

'It is a good river,' said Johnson.

After a while Wright said:

'Why did you want to leave us?'

Johnson's heart seemed to ring like a bell at this. He was touched by the delicacy and nearness of Wright, though he resented the intrusion.

'I wanted,' he said, 'to try this along because my father came this way. I wanted to see.'

Johnson was too simple to notice how this new motive had displaced the old haunting one.

'I would like to know what happened to my father.'

'He was a good way beyond this.'

'Yes.'

Wright saw the father in the thick-shouldered, shaggy-haired figure of the son. He respected Johnson's motive, but he was instinctively shy of investigating such a curiosity any more. To Wright the emotions must be protected by convention. The rippling shadows of the trees and the strips of light between them passed under the boat like a silent moving cloth.

The first shot was Wright's. It sounded like a dropped plank and its echoes went hard against the trees and leapt back. The birds rose black in thousands against the sky. Swiftly Johnson paddled to the fallen bird before it sank. Time passed and Johnson got his shot. An excitement possessed them both. The trees had given place to a scattered scrub which grew thicker in the distance. The rays of the sun lengthened in it. They chose a good landmark, tied up the boat and went ashore. The mosquitoes and flies clouded round their hats.

They had landed at the entrance of a densely overgrown creek and were walking along the thick bush of the bank above it. They saw the droppings of four-footed animals. They trod down a procession of great ants. The land smelled dry and pungent and clean to the nostrils. The grasses were browned by the sun.

A mile up the creek the banks were lower and the water had dried out of it leaving only a bed of caked mud pocked with holes of dirty water. This water was often alive with movement whirling round and bubbling like a simmering stew. Wright cut a stick and sharpened it, saying there would be fish in these potholes left behind by the drying water of the creek, and they went down into the mud. The movement of the water was made by the whirling of innumerable small electric eels but beneath them were fish. Johnson watched Wright stabbing the pool with his spear. 'Mustn't,' thought Wright. 'Mustn't stir up trouble.' There was no sun in the bed of the creek. Johnson still carried his gun but Wright's lay on the bank. They were too engrossed, concentrated on the luck of each dip into the black water, to speak or to notice any other sounds and the failing of the sun.

Presently a scattering of birds and scampering of feet in the bushes twenty yards away where the creek-bed gave a sharp bend and went out of sight, made Johnson turn. An extraordinary movement of alarm was in the creek. Johnson moved to the firmer high bank alert for what was happening there. He thought the noise might be caused by wild pigs. He whispered quietly to Wright and went four or five yards nearer to the bend and was standing waist-deep in the bush. And now the confusion and rustling alarm the flying up of birds spread down the opposite bank of the creek towards them. Wright stood shin-deep in the mud, looked up when Johnson called. He turned round to step out of the mud instinctively going for his gun. As his back was turned, the tall grasses on the bank opposite to him were broken down and the paws and head and shoulders of a jaguar appeared. It pulled up noiselessly at the bank's edge. In the grass its head was soft and marked with the greyish golden dustiness of an enormous moth; as suddenly and softly as the whirr of a moth the animal had appeared. It stood still, amazed, one paw on the top of the bank and one raised cat-fashion in wonder, arrested in its intent of running down the bank to its drinking place. Its eyes were like pits of gleaming honey. They had not seen Johnson.

'Tiger!' shouted Johnson. 'Keep still! Don't shoot!' 'Get the hell out of this,' he yelled at the creature. With light guns like theirs the only hope was to startle the beast away. The jaguar had been considering in these seconds the figure of Wright heaving himself by a bush-stump out of the creek-bed. The air popped out of the pits of his heel-marks. He looked like a scrambling animal, though he was unaware of his danger until Johnson called. The shout from Johnson startled the creature. It gave a swift turn of the head, crouching as if to leap upon the new voice, and then in panic swept round in the breaking grasses to rush swiftly away. Johnson's gun was raised by instinct though his shot was too light for such an animal. He knew it was fatal to fire and wound. But now he could hear the beast breaking the bushes in flight he jumped into the mud and scrambled up the opposite bank to get a sight of it. Wright now aware of his escape shouted, 'Don't shoot!' in his turn. Excitedly he was picking up his gun and turning to warn Johnson as he did so. Johnson stopped. His gun was pushed over the bank and he himself was half-lying on it, struggling to get up. In his hurry the gun went off and Wright shouted. Johnson lay still for a second in consternation at his accidental shot and then he realised that it was not his gun that had fired. He turned round and slithered comically down the bank, the dust pouring on to his head and shoulders, the thorns cutting his hands. He leaned staring against the bank at the end of his slide. There he saw Wright lying face downwards over his gun. Johnson blinked his eyes wondering why Wright was lying down. 'Why is he lying down to sleep? Is he tired?' Then he saw the nails of Wright's right boot and the toe twisted under a loop of root. Then the blood from his chest spreading under the arm of his khaki tunic.

Johnson crossed over to him and knelt beside him.

'Wright, what's happened?'

There was no movement and no reply. Carefully he turned Wright over and as he turned a murmur came from Wright's open mouth and the eyes quivered. His face was bloodless, red only in the faint fine veins on the cheek bones.

Johnson had no knowledge of what ought to be done for a man in Wright's case. His mind was a chaos. He undid the coat and saw now the burned hole in it and the tear in the blood-soaked cotton shirt. The charge had evidently entered the lung and Wright's faint breath was stertorous. Flies, the blown motuca, came in dozens at the smell of blood. They settled thickly on Wright's still face if Johnson for a moment ceased to drive them off. There was no drinkable water in the creek, only a black ooze, and neither carried any brandy. Johnson took off his coat and his shirt and the biting flies blackened on his bare skin, humming and whining, blowing into his face like a stinging, humming grit, as he tore his shirt to get long strips for bandages. He sickened as he wiped the wound and contrived to bind the strips round Wright's chest, putting a pad on the wound to staunch the thick drip of the blood. Wright's eyes opened in the middle of this and his lips moved twisting with pain. 'What happened?' Johnson said.

Wright could not answer.

To carry Wright on to the bank and leave him lying there to be tormented by the flies and by thirst and in the darkness a prey to any animal, while he got help? It would take two hours and he might be dead. The pulse was not strong. To carry him down to the river? That was a mile, a rough mile. The camp was a good three miles away, yet perhaps it was nearer across the bush. Hoping that in the quietness of the evening the sounds of shooting might attract attention, he went up the bank and fired ten shots in quick succession. He had only three left. But when the quietness had settled down again after the shots, the futility of the signal left him in despair. He was frightened by the silence. He shouted, knowing too that that was futile. He remembered his father had died in this country.

Wright moaned below on the bank.

All that anxiety to know how: to reconstruct what had happened in those already hazed seconds when the jaguar had appeared and then fled, fought in Johnson's mind with this picture of his father's death and the agony of not knowing what to do. He went down the bank. Wright's eyes were still open. His lips tried to speak. His breath when it came roared like gas in a burner.

'I'm going to get you up to the top. Can you move?' There was no answer but a closing of the eyes.

The evening sky was becoming green and darker, the bush soundless and black. It seemed to Johnson he must get Wright down to the river where there was water and the boat. But when he put his arms under Wright, he could not move him. Three times he tried and the sweat poured down his face and chest. He was maddened by the flies. Then a brutality came into him and, cursing, he put his arms round the drunk, will-less body and lugged it up. Stumbling, falling, sprawling on top of Wright, straining until he felt his heart and stomach would burst, he got him half-way up the bank. There was a clear way here and he wedged Wright's feet against a bush. Wright's arms moved in agony. Johnson sat there gasping, swallowing his sweat, looking down like a hunted animal upon the wounded man, with pity and ferocity. There enters with the handling of the sick a kind of hatred, a rising of life to repel the assault of evil.

'The poor bloody fellow. The poor bloody fellow,' gasped . Johnson.

Then once more he struggled till he got Wright over his shoulder and tottered with him to the top. The blood came on to Johnson's skin.

The stars had not yet appeared and this night the moon was late in rising. The one pleasure of running through the rough and broken mile to the river's edge was the freedom from the flies. Bats were flying out of the bushes and the moths were tossing over the thorns. Johnson ran. He was exhausted when he got to the shore and lay breathless for a moment. Then he pulled in the boat and sluiced his hot body and his head with water from the bailing can. He filled it with water and wedging his hat over it to stop the water from spilling, he went back. He could not run now because of the water. But now in the dark the country was so changed that it was hard to find his way. He began to think he had gone too far and wandered back. He shouted. He turned again and at last the moans of Wright brought him to the place.

I must get you moved before it is dark. I'll move you soon. Can you hear me? We'll soon be moving.'

The water had revived Wright. He looked into Johnson's face and nodded.

What shall I tell them if he dies? What shall I say to his wife and to Lucy? It is my fault, coming up this river. No, it might have happened anyway. It was an accident. What was he doing? I didn't see. I was half-way up when I heard a shot. On what river? That was not the river you were going by. Why were you on the wrong river? My father died in this country. He went by this place. He might have died in this very place. No one knows where he died. The Indians come here. There are fires of Indians tonight and no bloody moon. If it could have happened on a moonlight night. If I had been up further, this would not have happened, he wouldn't have found me today. This is Lucy. This is the ruin Lucy has brought on me. No, it was an accident...

'Can you put an arm round my shoulder? I say, can you put an arm round?' He hasn't strength in his arm. Shall we stay here? Shall I light a fire and the others are bound to come if we do not go back. How is it? He can't say anything.

There's a stupidity in the pitiable helplessness of the wounded. Wright moaned.

It is better if he moans. The flies have done. I wonder where the tiger is.

He went down to the creek, into the strange place which was nearly dark now, empty and without sound, where less than an hour before they had been poking in the mud-holes. A fish Wright had speared lay by the guns. The scene was not to be believed. Johnson found himself picking up the dead fish and bringing it back with the guns. He and Wright had seen it flap under the stick but had not even glanced to see it die.

Johnson hated the sight of the two idle guns now and they encumbered him; but he grimly made up his mind that if it killed him and it killed Wright he must carry Wright down to the river. If they waited Wright, for all he knew might die. He remade the bandages. The bleeding, he thought, had slackened. As he was putting on his coat Wright spoke and Johnson dropped to his knees to hear.

'Come here . . .' the voice faded.

'I am here. It's Harry. You're all right. I'm here. I'm going to get you down to the river.'

(The wrong river.)

'Lucy . . .' said Wright.

'It's me, Harry. Not Lucy,' said Johnson.

God, he's dying. He's dying and he's talking about Lucy, telling me he knows about Lucy. Would you deceive a man who is dying? Johnson knelt, with his face close to Wright's. The eyes were closed as if he were asleep and he stopped speaking.

I must get him back, dead or alive. I must carry him. Somehow he propped up Wright's body and, kneeling, got it on his shoulder, grasping him by the legs. He was strong now. He staggered up and stumbled forward in the thickening darkness under the first stars. He stumbled over roots, he tore his clothes on bushes, fanatically he followed the familiar bush of the creek bank. His shoulders were aching, his tongue out of his open mouth sucked in his sweat. Twice he rested and groped in the bush for sight or sound of the creek.

The stars were brilliant and clear. They shone with miraculous clarity, mapped clearly in their constellations. They placed a definite order before the eyes and one walked in the most marked and munificent light. But this order was in contrast to the confusion of the bush. Each tree where it touched the sky was like a bunch of black spears—each bush, each mass of grasses had this marked black head, clear and dramatic. And a voice seemed to come out of it, saying, 'This is the way. You remember this bush, and then the five trees together and the scrub you skirted. You counted the bends and the rises.' Each one stood distinct and black and certain. Johnson hesitated. Crouching under the groaning man, he turned round. Behind him, as before him, was the same array of definite shapes, a multitude of motionless caped figures. He swung round, but it was the same on either side of him. The definite things near by, the stars like tears in the branches, cold and brilliant, the heavens immaculate and lucid in their complexity. He listened for the sound of the river. There was no sound. The sweat went cold in his body. He lowered Wright gently to the ground and, turning with superstition at every pace to keep him in sight, stepped into the gap in the scrub where the creek was. He put his hat down on the gap and walked through.

There was no creek. There were twenty yards of low grass

and rock with stones shining in the starlight and then a bank of
scrub. This must be the creek. Carefully observing every step,
he went to the bank, which was a foot or two higher than the
land around him. There was no creek. He saw nothing, no line
of bushes which he and Wright had appeared to follow hours
earlier. he felt he had been lifted up and taken into country he
had never seen before. From where he stood he could see his hat
in the gap and he returned to it rapidly, dreading that it would
vanish or change before he reached it. He got there. His hand
trembling as he took it and now he made for Wright. The
world had opened loneliness upon him. In every direction it
seemed certain that the river lay. The dark bush did not lose the
distinctness, the simplicity of its shapes. He looked down upon
the pale face of Wright.

'God, I'm a bloody fool,' Johnson said. 'How have I done
this?' He stood stiff, ripples of coldness passing through his
body, unable to decide anything. Once he thought he heard the
sound of the river but it was a movement of night breeze
soughing in the distant trees and passing over them like some
lost human breathing. He fought with all his slow will the
impulse to dash here or there following this certainty and the
other.

'Wait. Wait,' he said.

He knelt down beside Wright and talked to him.

'Don't worry,' he said. 'We'll soon be there. I'll get you down
somehow. Just taking a breather.'

Wright's breathing seemed easier. The pulse was unchanged.
The ground, Harry felt, was sodden with dew.

But this trick of calming and distracting himself and then of
looking up with an open mind, to see the scene afresh and find
conviction in a flash, did not succeed. There was a momentary
illusion of vision, then it dissolved.

He thought, 'Shall I pray?'

If there were someone outside or above, with simple ease this
person could point out the path to the river. It would be very
simple. He thought, 'Our Father which art in Heaven'; no,
what's the use of panicking! I'm not going to wander round in

circles. Light a fire. They'll see that. It can't be very long before they start searching.'

He said, 'This is not admitting defeat,' and 'Wright will not die.' There is always some other small thing and after that another small thing which can be done before a man dies.

Harry had no watch. He stood trying to calculate the time. He set about lighting a fire, gathering the dry sticks. He took out his matches and his pipe and put it in his mouth. Then he could not, for some reason, smoke while Wright was lying there helpless. He lit the fire and as its light made a glowing room from which the sparks dances, his spirits rose. He did not like the thought of the black, distinct caped figures of the trees behind that unnatural and fluid wall of light. He worked hard collecting and piling on the sticks. The flame went up in a waving spire. He worked ceaselessly, taking no notice of Wright. 'It was an accident. It was an accident.' Branch after branch he brought and made a stack within reach of the fire. His whole life went into making the pile. The fire blazed high, yellow and dancing. Like an animal leaping, some yellow cat, the flame licked up in the dark, sending out claws at the darkness. He looked up and he and Wright seemed to be in a huge glowing temple, higher than the highest trees, wide and palatial. A fire that could be seen for miles. He shouted and listened. There was no answer, yet there had seemed to be a thousand faint answers, the movements of leaves or the scuttlings of night animals. He sat down beside Wright, exhausted, his throat dry; realised now how his head ached and that he was sick with hunger.

But Johnson could not sit by the fire and wait. 'In a moment,' he said, 'I will go and look for the river. If I make the fire high it will guide me. I can't be lost. I can get water for him.'

Wright was murmuring again for water.

'Poor devil. I'll bring it you. Just getting a breather.' The heat of the fire was strong and flat against his skin. He stared, exhausted, thinking out his plans.

And suddenly it was curiously easy, as if in the darkness a hand behind him guided him through the scrub. And it was near. Nearer than he would have imagined. They were camped, he discovered, within fifty yards—no, it seemed only twenty—from the river. The creek bank and its long clump were just as he remembered. He went down the bank of the river and so great was his joy that he did not even look for signs of the rescuers, but himself put down his hat—now the only thing he had for water— beside him and drank deeply from the river. It was cold and glorious water, so cold that it made his hot lips and his dry mouth sparkle with delight and his body shuddered.

Shuddering, in amazement, he woke up. The fire had gone low. He had dreamed.

He could not tell how long he had slept; it seemed only a few seconds; yet the lowness of the fire showed that it must have been much longer.

'Wright!' he called. 'Wright!'

'Thank God,' he said when Wright murmured.

Johnson could not see his face. He jumped up and put more sticks on the fire. His head was throbbing violently. The flame started at once, but now the glow was weary and wretched. His body ached. He heaped on the sticks.

'Good God,' he said. 'Where are they?'

He shouted to no answer.

'I must go now,' he said.

He turned to shout again and then he saw he could not go.

Seated like a large dog at the rim of the circle of light, pale and dabbled by it and unmoving, was the jaguar. The animal had, indeed, been many yards nearer when the fire was low, but had turned back when Johnson had awoken and made it up.

Days afterwards, when Johnson could speak to Phillips of the happenings on this night, he said:

'It was the most bloody awful luck that we hadn't taken the rifle. I damn and curse myself for being such a fool. We shan't see another and he certainly won't come to sit and watch us, like that one did, as if we were a pair of clowns in a circus ring.'

Johnson stood still with the branch in his halted hand frowning at the jaguar. Like nearly all animals they are, he knew, afraid of man and avoid him, but there is a point at which fear becomes fascination. If he stepped out of the exorcising circle of firelight and walked out into the dark in search of the river, the animal might recover from his trance and follow him.

'What do you want?' called Johnson sharply.

The creature pricked its ears.

'Clear off,' Johnson shouted.

The jaguar rose and moved nervously away, but fascinated by the fire, did no more than move further round the circle. Once more he sat down like a dog with heavy front paws. Johnson knew he was safe with the fire. His real concern, amounting to a shocked anger with himself, was that he had fallen asleep; his only fear that he had slept for hours and that the camp party had not seen the glow of his fire because it was low. They might have passed hours before.

The life of Wright was the important thing. Harry picked up his gun and fired another shot. The echoes fell in a hard rebounding shower over the bush. The jaguar started up and crashed away into the darkness. There was the old silence swirling into stillness like a dark pond after a stone has been thrown into it, and the rim of the circle of light had a more sinister loneliness now that its sentinel had gone. It was not a time when it is easy to be patient. One counts the minutes.

There was no answering shot.

Johnson turned to Wright. His lips were cracked and blood-less, his tongue protruding and dry, his eyes staring. He murmured sometimes words and names which Johnson could not catch.

'Why the hell don't they come!' said Johnson. 'Has he got to lie here and die because those fools don't answer?'

He stood up and fired again.

'They're coming,' he said to Wright as the echoes rained. But he had no evidence that they were coming.

He cursed them quietly; but he was still most appalled by his own guilt in losing his way and in falling asleep. He was eaten by shame and by horror at himself, his ignorance, his incom-petence and his guilt. He walked up and down looking at Wright, maddened by his inability to do anything. In his mind he continually saw a brilliantly lighted room—the drawing-room of Wright's house in England— and there Mrs Wright was reading and Lucy was standing by the open window. They were talking. Suddenly he was there walking across the room and they got up and walked quickly, exclaiming, towards him. They came very close to him, Lucy was laughing and the laughter and some words passed near to his face and then over and beyond him, and once more the room reappeared as it had been at first, with Mrs Wright and her book and Lucy at the open windows. They got up and came to him as before. Over and over again, with the tireless mechanism of pictures, these two scenes were enacted. He could not shake them out of his mind, as he bent to collect sticks or piled them on the fire or turned round to speak, for his own relief, to Wright. Johnson had never seen a dying man before.

He stood looking up at the darkness and the words came to him, 'My luck has gone.' He saw in the darkness the lighted room, the two women, and then beyond them hundreds of small fragments, glittering, out of his own life and his father like a shadow thrown upon it all. He felt again what he had felt intermittently during the past six months, that he had no longer a self, that he was scattered, disintegrated—nothing.

Then the jaguar returned and sat down in the rim of light and its eyes were as brilliant as motionless lamps.

'God!' exclaimed Johnson and, without thought, advanced running towards the creature with a burning root in his hand, shouting, 'Get out, you fool! I'll beat your brains out if you don't clear out.' He raised the torch of shrub high as he ran, shouting. The animal turned tail and sprang before him into the scrub breaking down the branches, and Johnson went after it. For fifty yards he ran and roosting birds clapped up in the dark. The ground rose and he heard the tiger still springing far away. And then Johnson dropped his arm in amazement. There was the river streaming in the rising moon within a hundred and fifty yards of the camp. There was the creek bank. He looked back. The camp which had seemed to be on rising ground was in a wide hollow and its light was invisible. The river was exactly in the direction from which he thought he and Wright had come after the accident.

He ran back to the camp. He marked the direction by the brand; and with rough care for Wright he knelt down and got him on his back. His weight was dead. Staggering with the man he made in the direction of the river. Wright groaned as he jolted over the rough ground.

There was no sign of the jaguar, no answer to his shout as he stood on the shore.

He paddled out to midstream to be in the path of the rising moon. 'Then it must be nearly midnight.' No action or sensation of Johnson was nervously harassed or feverish. His struggle with the weight of Wright, his staggering blindly through the bush, his guilt, his visions of Wright's home and of his own life, culminating in the words, 'My luck has gone,' he experienced slowly and laboriously. He passed through this suffering like an ox.

The current was with him. On the blank surface of the air were scratched the thin night-piping and croaking of water birds, but as the moon came up the surface began to glimmer. Faintly at first his shadow and the shadow of the gunwale were placed like hands upon the form of Wright and his face took on a deeper waxen whiteness.

'You're all right now. We're there,' Johnson said. The warm wash of moonlight unclosed into a radiance rich like the whiteness of a lily and the river became like a white path of voluptuous funereal marble between cypresses in some southern cemetery. The night was warm.

All Johnson's thoughts were fixed on the camp, estimating the distance, noting landmarks, his eyes constantly searching for the gleam of the fire. Not for one moment did he think, 'This is the end of the expedition,' but he thought of the journey back to Calcott's town and who would take Wright there. To him every one of his paddle-strokes was something that detained Wright from dying. He was confident of his judgement though his luck had gone. His anxiety was that the others, who had not apparently come out to look for him, should have let their fire go out.

Presently, far ahead of him, he heard a shot. It came from far down the river and, seemingly, from the opposite bank to the one where the camp was. He took his gun and fired an answer to it. An answer quickly came. He paddled rapidly.

'They're here,' he said.

Wright began to gasp and rave and then fell quiet. Where the hell are they? What are they doing down there?

A strong smell of burning wood hung over the river, dry and acrid. It blew over from the opposite bank. He passed—he remembered it—the opening of a wide creek—and suddenly voices were plain. They were coming from the creek. Loudly another shot sounded. It was from the creek. He paused and shouted. He shouted several times. The voices came confusedly over the water and then there was an answering shout. He turned the canoe towards the sound and as he approached the creek mouth he saw their boat come down. Again he shouted and now there was no doubt about it. They called, and from under the fantastic shadows of the branches, the men rowing in the bows, the black craft appeared with Phillips and Silva standing in it. Then there's no one in the camp. The fools. Any animal may have pinched the stores.

They came alongside.

'Don't run me down,' Johnson said. 'There's been an accident. It's Wright. He's got shot.'

Phillips and Silva looked down into the canoe. 'We've been searching for you. The trees were fired opposite and we thought you were up the creek, cut off.'

'Don't move him. But let's get to the camp quickly. Where is it?'

The men in the boat were silent. The Brazilians gazed down at the figure of Wright. 'He's dead,' they said among themselves. Feverishly they rowed over and Johnson went ahead of them, the two parties shouting across the water.

They arrived as Johnson was pulling his canoe into the shore. They jumped into the water, ignoring their boat to crowd round the canoe.

'Look after the boat,' Johnson said. Two went shouting after it into the current to tie it up.

'He's unconscious,' said Johnson. 'Lift him carefully. It's the chest. He's lost blood.'

Easily they lifted him ashore and laid him on their coats on the ground. They switched on their torches. The men were called to make up the fire. Johnson and Phillips knelt beside him and Silva was opening the medicine-box.

'Harry,' said Phillips in a startled voice. 'He's not unconscious. He's dead.'

They both stared at the white face, the staring eyes, the protruding neck. 'He's alive. He was speaking in the canoe.' But when they felt the pulse and listened for the heart and put a mirror to his lips, they knew he was dead.

They stood up and all gathered round. Their torches played in balls of light about their feet and they stood in the vivid whiteness of the moon, looking speechlessly into one another's faces.

CHAPTER ELEVEN

ALL night their voices went on. They talked to create the illusion that he was still alive. The moment one voice stopped another began so as not to leave a silence in which his death could become real to them. They had spread a mosquito net over him, and that suggested he was no more than ill and sleeping. Themselves, they had a horror of sleep.

But sleep overcame them. They sank down where they were before dawn and Silva covered them. So deeply did they sleep that they were not awakened by the crew, who came noisily to look at the body under the net. The sun was high when they woke up at last. It was a shock to wake up in the face of the unchanged sun.

Even when they set to work with two of the crew to cut out a grave in the root-tangled earth, they could not believe he was dead. There were no spades. They had to use their long knives. They unbent from their task and, looking down to the river, expected to see him stripped and washing there, waited for him to come and laugh at them for sweating at a hole in the earth. They expected he would stand there and ask them what the hell they thought they were doing.

Phillips was ill. Twice he had to leave the job and when the time came—which they all dreaded—to lift the body, he got away. Johnson beckoned to Silva and those two carried the body to the grave. The crew stood by watching. When the body was put into the trench, the crew crossed themselves and then a murmur started among them.

'What is it?' Johnson said.

The men were protesting that Wright was being buried in his boots. These had been admired. Johnson told the men to shut up. He had them stand in silence at the graveside. Phillips came up and looked away at the sky but the others were watching Johnson. He was the leader.

'All right,' he said after the silence.

Then Silva, who had made a cross out of two branches of wood, stuck it at the head of the grave. Phillips and Johnson looked with surprise at him. They had not thought of this. Afterwards they were grateful.

Johnson went to his hammock. They were all hungry, but except the men, could not eat. They stood about in the camp pretending to do things but chiefly stopping to stare at the river. Only the men watched them and talked continuously, sitting close together in the boat.

'What are they saying, Silva?' Phillips asked.

'It is the boots. And their pay.'

Phillips said to Johnson, 'They'll dig him up for those boots.'

Johnson had spoken very little, and he seemed to add this remark, in his quietly listening manner, to the stock of his thoughts. Gilbert above all wanted to talk. To discuss again Harry's account of how it happened, to talk of death and Wright's life, writing to his family, the future of the expedition. He assumed that all this was Harry's responsibility. Harry did not talk. A feeling arose in the camp that Johnson must take responsibility for Wright's death. He lay apart and everyone watched him. The smoke of his pipe rising up from the hammock showed them he was at least not sleeping.

If by chance they came together, they were conscious that Wright ought to be among them. Gilbert thought, 'That is what might have happened to Johnson's father—an accident with a gun.'

As the vacant day passed among too much greenness the huddled crew grew sullen and noisy. The lack of drama in the death and the burial, the absence of some outlet for their feeling about death, gave them a blunt anger. If Gilbert gave them an order, they lowered their eyes. Presently their murmuring grew louder. Silva awoke from his diffident, passionless trance; he had heard the word 'murder.' He mused about this. Then he came and sat beside Phillips.

'It is possible,' he said, 'that this was not an accident. Perhaps he killed Wright.'

'What the devil do you mean?'

'Perhaps.'

'Is that what they're saying?' said Gilbert.

'Yes.'

Gilbert, hungry and yet without appetite, felt sick and looked with disgust at Silva.

Silva shrugged his shoulders.

On the previous night when they had gone to search for Harry they had turned back to the wide creek where he had ultimately seen them. They had heard loud reports which had sounded like gun shots and when they got nearer they had discovered that the forest had been fired. Perhaps Indians watching the camp had fired the bush.

It was Silva, theatrically carried away by the thought of fire and Johnson and Wright perhaps trapped by it, who had insisted on going up the creek. The fire was not at the water's edge, but the reports of the burning wood, the wind roar and hum of the ascending flame, were loud. They had drowned the distant sound of Johnson's signals. And then Silva's imagined picture had proved to be wrong—perhaps fatal. Phillips blamed himself and blaming, despised Silva.

But it was the peculiar faculty of Silva to infect others with suggestion from his fantasies. He had infected the easy Calcott, he had gained the confidence and awakened the imagination of Johnson—now, once more, he insinuated doubt into the ready imagination of Phillips.

Phillips did not believe him, nevertheless the suggestion remained.

'He could have killed him,' said Silva. 'He was in love with Mr Wright's wife. He told me. That is why he ran away.'

Phillips listened with astonishment.

'Don't talk nonsense. Mrs Wright is a woman of fifty.'

Phillips saw that Silva had mistaken Lucy for Mrs Wright.

'When she had a baby Mr Wright would know.'

Phillips smiled at Silva's story.

'When did he tell you all this?' he said.

'He told me,' Silva waved back down the river.

Gilbert's *amour propre* was injured because Harry had confided in Silva and not in him.

'He was ill,' said Silva. 'He would not speak and then he told me. I told him about my wife.'

'You're not married?'

'Yes. She is blind. She lives with her family in Portugal.' The misfortune of Silva gave him seriousness.

'Listen,' said Gilbert. 'It's all nonsense about Mr Johnson killing Mr Wright. We don't do those things and he was his greatest friend and he is engaged to be married to Mr Wright's daughter. Look,' said Gilbert, as if talking to a child. He pulled a note-book out of his pocket. It was his diary. 'I've written down exactly what happened. You are to tell the men and stop this. D'you see?'

'All right,' said Silva dispassionately.

'Why,' said Gilbert, 'Mr Johnson did his best to save Mr Wright's life. We shall be going back tomorrow, I expect. You can tell the men that.'

When the chance came, Gilbert said to Harry:

'It would be a good thing to write down your story because we shall have to report this when we get back.'

Harry said that he had not thought of that. What was he going to write to Mrs Wright? And to Lucy? When was he going to write? Now, or in a fortnight's time when they got back to Calcott's?

Harry blinked in bewilderment.

Phillips saw that he had not been thinking of these things. Phillips was moved by the state of Johnson. The death of Wright had frightened Phillips, it had shrivelled his heart, made him look with apprehension at the country, with nausea at the changeless brilliance of the sunlight on the river and metalled on the trees. But he saw that Johnson was moved by sorrow for Wright's death and Phillips was suddenly shaken by his own lack of grief, his concern only with himself. Yet he could feel, beneath the shining glacial surface of his mind, the movement of grief. He grieved rather for Wright's wife and for Lucy.

He said to Johnson:

'It was not your fault. You did all you could. A wound like that—it's a wonder you got him as far.'

(Only much later on, weeks later, did the story of the second appearance of the jaguar accidentally escape from Johnson.)

'It was my fault,' Johnson said, 'because I came up here.'

'An accident could happen anywhere,' said Phillips.

They felt the absence of Wright on this first night.

It was a bad night. The stars were wiped out between two and three in the morning by a black cloud and lightning sheeted the river and lit flares in the trees, but there was only one abortive crack of thunder, a peal split and cut short like wood splitting. But rain came down straight and dense, hissing in the fire and soaking them, and the wind moved like a heavy sea in the trees. The rain washed away some of the earth from Wright's grave, making his body nearer as if he were trying to return to them, and uprooted the cross. Hurriedly in the morning they piled on the earth once more when they discovered this, and Johnson made them collect stones and build a low cairn on top. The work made them forget their soaking and the steam rising from their soaked equipment; and the physical labour gave them an escape from their horror.

With the job done, a feeling of expectancy arose. The feeling of suspicion against Johnson seemed to have subsided. Each man was concerned with himself, drying his things and thinking of the future, although when Phillips and Johnson came together they would pause and both feel the incredible absence of Wright. They had lost their sense of time. Their minds were always drifting back and living in the days they had been with him and had heard his voice and had seen him walk along the shore or step into the boat or dip his bare arms into the water. They remembered his extraordinary youthfulness, his way of treating the country as a playing field.

It was strange to them that the sun still rose and the days continued, that time went on. And it seemed that the country and themselves lived in different periods of time, they in the past days with Wright and particularly in the day when Wright had landed at the camp. When the men looked at them from the fire where they were cooking, they too seemed to be in a different period of time.

And Silva stood half-way between them, belonging to neither period completely.

The death of Wright had done one thing which, much later on, was to have its consequences. It had broken the intimacy between Johnson and Silva. Silva, on his side, had his new fantasy. Johnson, on his, saw in Silva the figure of his guilt, since Silva had been his companion in the flight; but the main reason for the ceasing of the intimacy was the simple one that misfortune brought the two Englishmen together because they were Englishmen and left Silva with his own people. The two Englishmen were too dazed by the tragedy to observe that this had happened, but Silva had observed it and, since it was natural, it seemed admirable to him. His companionship with Johnson had come to an end. He had no malice, no disappointment, no disillusion. It was like a work of art in the rough, essentially finished but which now needs the refinement of meditation, the suppressing of some details, the elaboration of others. Already he saw himself at home, telling the story. And since he was not at home yet, he was impatient and bored, waiting for the order to pack up and depart.

After the cairn had been built, Gilbert said to Harry:

'When are we going back?'

Harry said:

'Do you want to go back?'

The shock of this question showed Gilbert how deeply Wright's death had made him long for safety and return, how acutely he felt that to go on now was to be unprotected and vulnerable. They were like a holed vessel.

'We can't go on. The men won't go.' Gilbert could not tell Harry of the word 'murder' but he said: 'They think this is bad luck.'

'They think I have brought them bad luck.'

'We know that's nonsense, but . . . go on where? Back to the river?'

Gilbert knew that Wright would not have stopped because

one of them had died. Harry and Wright were the same. They did not think of others or of themselves, but of their purpose.

'You mean,' said Gilbert, 'you mean that we must go on with his job.'

Gilbert was to have a shock. He had been so used to feeling his own isolation from the traditions which Johnson and Wright took for granted, that he had argued that only himself acted individually. Harry said:

'No, there is no point in that. The men won't go. We must send them back.'

All this, Gilbert saw, was difficult for Johnson to say. Gilbert knew him well enought to know that he disliked giving reasons for his actions, but this was obviously not all of Johnson's difficulty now. The obstinate look was on his face, the visionary light was in his half-lowered eyes, the distant preoccupation with a purpose he was set on and which he could abandon only if he had to, with regret.

'I mean we will go up this river. I wanted to go up this one.'

'Change the whole plan?'

'Yes.'

'And the men?'

'Send the men back, I said.'

'Oh yes, I forgot.' Gilbert was bewildered, startled by the definiteness, the decision of Harry's purpose. Harry seemed not to be suggesting, but to be deciding.

'You mean,' Gilbert said more hopefully, 'just go up for a few days until you get to—where were you proposing to go? A week up the river and then back?'

'No. Go up there. Not come back this way. Go across,' Harry said vaguely.

'With Silva and me? Silva won't go. He's bored. He wants to go back.'

'I don't want Silva.'

They spoke in lowered voices, murmuring and whispering; if one raised his voice in the discussion, the other glanced at him pleadingly and the voice was lowered. Wright was too near.

'But we can't go on alone. No maps. No guide. And what about food?'

'Live on the country. There was never any need for all this stuff.' He pointed to the cases and bags in the camp. 'It hampers you. You can't move. You're tied to the crew and the river. We've got Wright's map—it isn't much good. It's been no good so far. This river wasn't marked on it. Still, if you want a map I've got one. I've studied this. I know the way I want to go.'

Gilbert argued that the planned expedition would be better than an unplanned, casual escape into the wild like this. 'And without an object,' he said emphatically.

He was incredulous. He was thinking too of the Indians. This wild proposal of Harry's would mean the death of them if they were fools enough to undertake it. He knew Johnson's reckless-ness, how out of periods of torpor and laborious, lazy, routine plodding it woke up; he had seen it in England when sailing—he refused to go out more than a mile or two on a fair fine day and then when half a gale was blowing, he was restlessly standing at the window, thinking, pretending to agree with people that it was impossible, and at last slipping out on his own.

'I've got an object,' Harry said. And now the resolute sullenness appeared to thicken his shoulders and put weight in his stance. 'I want to chuck all this and find out about my father. He came up here and I know all about the way he went. I've been studying it.' He hesitated and added with candour:

'I wouldn't care to do Wright's show now, though it looks like letting him down. It was my fault he died.'

The illogicality did not strike Phillips at the moment but afterwards he thought: 'If he wishes to make amends why doesn't he propose to finish the job Wright had set his heart on?'

But Gilbert was so distressed by the prospect of going on at all, he had, in the panic caused by death, so set his heart on returning, that he argued forcibly that Wright's plan, if any, was the only one: and the more he argued with Harry about it, the clearer it became that Harry was possessed by the flight impulse again. Yet it did seem to Gilbert that there was a kind of respect to Wright in continuing up this river. If Harry's flight up this, the wrong river, could be construed as a contributory cause of Wright's death, then you could wipe out that by making this the right river. It made Wright's death on it in some way less futile.

Harry went out of the camp to the place where he had wept after he had been caught by Wright. He saw the trees and the wheel of the river bend as he had seen them at that time. His misery had been enough then but it seemed pitiably small and self-regarding beside the emotion he felt now. Then he had committed some muddled misdemeanour, sunk into an inexplicable humiliation; but now—the life-long, guilty sensation of being different from other men had landed on its whisper from year to year until it had culminated in crime. He had committed crime. He had murdered Wright.

You could argue you had not murdered him, that it was an accident, something coming haphazard out of the year, descending upon your shoulders by chance. You could argue that everyone feels this guilt towards the dead when a man or woman dies. You begin thinking of the things you could have done. You say, 'If I had done this or if I had only spoken thus,' and so on.

But if this is so; this is to say, if *everyone* feels this, why do they feel it? Why this universal guilt? Why this desire to say, 'I have committed a crime?' Why, 'I am in some obscure but ineradicable way, a murderer?' Why, with the instinctive assumption of this guilt, is there mingled the intense and singular fire of an utterly distinguishing pride?—as if one said, 'He's dead, I killed him but I am alive, far more alive because of his death, and my remorse is like coals thrown on to the fire of my life making it burn unquenchably'?

One has behaved badly; one has broken some not very important convention of one's society, one has had a child by a woman one cannot marry and one has broken an agreement with a leader because of it. It would not look well, one's pride would suffer, if one returned and this was said. And so he makes a life and death drama of it, magnifies the little muddle to the dimension of a great conflict. Out of a little fear is created a great God.

This is one aspect of your case but you feel it's too mean and local an interpretation of your life. The fear you have had is not small, it is not the product of the habits of the street you were born in: it is Fear itself, a great principle that feeds as it sows. You cannot face the criticism of your fellows, but this is because you have a remorse not merely for your breaches of their code but for the whole train of events issuing mysteriously from the lives of your parents and passing into you when they took their form. It is a remorse for living, and there is redemption only in resignation or in the pride of some surpassing pilgrimage.

The pilgrimage which surpasses all others is the one from which all wish to return has been withdrawn. 'I go unto Him that sent me.'

Uneasy in these days if Johnson left the camp, Phillips noted the way he went and followed him. Gilbert feared what might be in Harry's thoughts. He watched Harry for a long time, respecting his meditation; then his desire not to be alone became too strong. He went up to him.

But Harry could not or would not explain what had passed in his heart, and Gilbert was left to conclude that grief had strengthened Harry's obstinacy. If they were to go on, surely it must be on Wright's route. Convention demanded it. If you thought you had dishonoured Wright, then here was the obvious way of restitution. The alarming thing was that Gilbert knew Harry's heart to be overwhelmed, and yet he saw him now beyond the promptings of the heart. There was an air of 'Now I'm the leader and I'll do my own show,' which in some way was distasteful.

Gilbert worked to talk Harry back to reason. After all, one may die but one doesn't seek to die. And there way Lucy. Every time he thought of Wright the image of Lucy came into his mind. The image had broken through everything and stood pleading with him.

'But I will only go if we take the men,' said Phillips, in his exact harsh voice. 'They were certain it was Indians who fired the trees the other night.'

'They didn't touch us. They won't.'

'But if we were attacked . . .'

'It doesn't make any difference if there are two or seven of us. A lot of rot is talked about Indians.'

'It isn't rot at all. You know that as well I do. Its absurd; two men can't go for a country walk across God knows how many hundreds of miles of forest, where there are no towns, where there's no food and only savages. You can't do that. I refuse. I refuse to go or to let you. Unless you want to die.'

Johnson gazed at him gravely.

He did not appear to listen.

'My father did it,' he said.

'Yes, and he was killed. He was mad. Harry, my heart's not in this now Wright has gone. He wouldn't have let you do this. And I shan't. I'll do Wright's plan properly if you insist because I said I would in England, but not yours.'

But Harry, who had travelled alone before, though not in this part of the country, knew exactly the difficulties. He knew exactly that it was reckless, that it was almost impossible. The events of the last weeks, culminating in the tragedy, had made the impossible exquisite in its fascination.

'I don't think you need worry,' he said persuasively.

'Aren't you afraid?'

'I don't think about it much. There is always something to do.'

'But if the Indians took you, if you had no water, there's snake bite and everything that could happen.'

'There is generally something to do.'

He was impervious.

'Well, I don't know.' Gilbert paused and frowned, strained by doubt.

'No,' he said again emphatically, 'it's final. I'm sorry. I'm probably a coward. But I'm not going.'

'I think you would find that you take a mistaken view of yourself,' said Harry. 'You would find that it would suit you very well. Sensitive people think they are afraid when they are not really.'

They left each other and Gilbert went off alone. He sat by the river and the flies pestered in hundreds like a mist about his head.

He moved on from place to place tormented by them. He went off astonished by Harry's words about himself. He was astonished that Harry should even thing about him. He was curiously flattered.

But he was wretched. He had gone romantically out and he had failed. The crisis came, and he could not rise to it. 'It is like this in all my life. I back out. I drop low. I talk round it and get out of it.'

He thought that there are crises in one's life and that it is not until after they have passed and one has taken one's decision that one realises they were crises that have decided important things. Now it would have established for ever that in everything he was a coward—unfairly established.

When they were alone together Phillips said:

'When are you going to give the order?'

'What order?'

'Going back.'

'They can go tomorrow.'

'We, all of us?'

'No.'

'I've told you I'm not going.'

'That's all right,' said Harry. 'I understand your point of view. You would be surprised to see how good it was if you changed your mind. I'll go on.'

'Alone?'

'Yes, I think I'll go.'

Phillips' heart jumped. He looked at the reddened face of Johnson. He had grown a scrubby beard and his hair was clotted over his ears. The beard curled absurdly from his face. He looked fantastic in his seriousness.

'You can't go.'

'You can't stop me.'

'What about Lucy? You've no right to kill yourself. What about her? Do you want to die? You'll never come back.'

'Lucy,' said Harry, sulking suddenly, 'has no claim on me.'

'Yes she has.'

'What?' said Harry roughly.

'She loves you.'

'Leave her out of it.' Harry's roughness subsided. 'I shall be all right.'

'She's lost her father, you know.'

'This hasn't got anything to do with it. It's easy. It'll make us a month late, that's all.'

'It has everything to do with it,' said Gilbert.

'There have been no letters from her,' said Harry. 'It's all over.'

'There was no time for letters,' said Gilbert. 'I had no letters from England.'

'But Wright had.'

'Harry, you're run down and in no state to make up your mind about anything.'

'I can't see,' said Harry, appearing to weaken, 'what you've got against my going.'

'I don't want your life on my conscience. If anything happens to you I should never forgive myself.'

Then Harry said a revealing thing.

'I'll never forgive myself about Wright. He needn't have died.'

Gilbert could see under his determination the gnawing of this agony. He saw that it eclipsed everything else, distorting a whole vision of the world. And so now, as the talk continued, he humoured Harry and spoke with sympathy, trying to learn the whole of what was on Harry's mind. But shyness on Gilbert's part and the fact that Harry was so imperfectly expressive of his feelings, so that he was like a big animal who can only make meaningless grunts and moans, made revelation impossible.

Before daylight went Harry called him to the list of stores and ammunition and showed him the map and the notes he had made. He worked out the distances and the time. A week on the river and four weeks overland. The ground rose to flat-topped mountains and bush. 'If we discover anything certain about my father I promise you I will return this way and I am going to leave the boat. Hide it.' His meditative voice became eager, but he was indifferent to any of Phillips' arguments and talked to him in a friendly, polite way, casual but not pressing.

There was an immense distance between them. Phillips was convinced at last that Johnson was asking him because he felt it unjust to deny him the adventure and because there was nothing to do but ask him since he was there. Johnson was being polite and, really, patronising. Phillips' vanity rose up.

In the morning they packed, but still Phillips did not believe in Johnson's intention. Phillips worked hesitantly; he saw Johnson had divided the stores.

'Damn him, I can't leave him. He's mad. I shall have to go. I shall have to die too and I don't want to.'

'God!' exclaimed Phillips in a burst of rage. 'You're not going. I tell you you're not.'

He shouted out before all the others. The men stopped and stared. Johnson said gently:

'Don't be unreasonable, Gilbert. You're taking this all the wrong way.' He put down the sack he was tying.

'Silva, what do you say?' called Phillips.

Silva came up.

'I am not going,' he said.

'But he mustn't go, must he? Two people, one person, even the whole lot of us—it's impossible.'

Immediately Phillips felt foolish, sunk in cowardice. Silva shrugged his shoulders.

'The Indians will eat you,' he mocked with a gentle teasing sadness that was so pretty that it was ridiculous. Harry laughed and the men laughed too.

The men looked ironically and nonchalantly at Johnson. They were eager to be gone. They considered him as good as dead.

CHAPTER TWELVE

IN the kitchen of Calcott's house, in the brothels and bars down at the quays, women were saying that José Silva had come back. He was not dead. He had returned. He was a fool to trust the foreigners. There was no money. They had discovered no mines. José Silva had come back without a penny. There was a rumour that he was dead, but he had appeared, the miracle, shining like a ring.

'And the Englishmen?'

'They are still there. I left them. The journey was beginning to bore me.' He waved to the river.

The gendarmes poured a little dry tobacco in their hands and tilted it into their cigarette papers, listening. One of the Englishmen dead? They put their cagarettes into their mouths, lit them and looked at the forest breaking up to the river's edge, the desolate place where the rubber trade had once flourished. It surprised nobody that one of the Englishmen had died. Hundreds of people died, and very young. Hundreds also were born.

'The old one? With the beard?'

'Accidentally, with the gun.'

'Had they the permit for the gun?'

A point of law was more interesting.

Silva went to Calcott's house. Calcott received him with aggrieved sarcasm.

'Had a nice holiday?' asked Calcott. 'Been having a bit of a joy-ride, haven't you?'

But Calcott was overjoyed at Silva's return. He had been lonely and bad-tempered for weeks.

When Silva had gone, and then Wright and Phillips, Calcott had shut himself up for days, staring at the open Bible and drinking some whisky which he had removed from Wright's possessions. He had wandered about the place, sulking and querulous. He had had a serious quarrel at one of the brothels and had been turned out. When he came back he had beaten his wife. After this his emotion passed and he lived through the torpid weeks like one waking heavily from a dream.

'The missus has given me another little birthday present,' said Calcott to Silva, relenting, eager for friendship. 'A girl. That makes three girls. This country's no place for children. I'm going to take them to England. They've never seen England. That's funny when you come to think of it.'

He preferred the earlier bastards to the later ones born in wedlock.

Calcott took Silva to see his new child. Mrs Calcott was sitting on the doorstep in the late afternoon sun, suckling the child whose eyes were like big, gleaming apple pips. The mother was singing and her face brightened when she saw the man of her own race.

'Ay!' cried Silva with delight. 'The little beauty.' The other children gathered round the delighted, cooing, dancing little man. Presently he changed.

'Wait,' he said to Calcott. He patted his pockets. 'Come inside. I have letters.'

The letters were two: one from Johnson and one from Phillips. Silva, Phillips said, would tell him more fully about what had occurred. Johnson asked that Calcott should pay the men as arranged with the emergency sum which Wright had left. Wright had unfortunately shot himself and had died. They were sending Wright's effects, his letters and his money by Silva, to be sent to England, as they themselves were going on. There were letters for Mrs Wright. He was to send any letters to Rio. They thanked him. Phillip's letter gave a fuller account of the death of Wright.

'They're going to look for Mr Johnson's father,' Silva said. 'Mr Phillips did not want to go. He was very angry. I thought they were going to kill each other. Mr Phillips was afraid.'

'Bloody snob,' said Calcott. 'I thought he was yellow.'

'He said they would be killed by Indians,' Silva said.

'So they will,' said Calcott.

The Englishmen had become remote in his mind; only slowly was he recovering his memory of them. But when Silva described the journey, Johnson's flight and then the death of Wright, Calcott began to realise what had happened.

'How long have you been coming?' he asked.

'Three weeks,' said Silva.

'Three weeks,' he said. 'What's the good of that. The whole bloody issue's dead by now. Would they listen to me? Oh no! I'm nobody. I've lived in this country thirty years but I'm nobody, see? Not good enough. Not a bloody gentleman.'

Calcott cooled down and murmured to himself:

'The poor bloody old fellow. I thought he had more sense. Fancy coming out here to shoot yourself with your own gun.'

Again Silva had to tell the story of the death of Wright and answer Calcott's questions. He became gloomier.

'He was in this very room,' Calcott said. And he went on saying things of this kind, such as 'If it's meant, it's meant.' 'He didn't know when he said, "So long, ol' man," that it was the last. I've seen men go out of this room smiling and the next thing they're still. Eh, Silva?'

Calcott's eyeballs began to protrude and his manner became more dramatic as the death of Wright became clearer to him. He seemed to see it standing in the room.

'I'm sorry he's dead,' said Calcott. He stood up at the table. 'I'm damn sorry. He was a gentleman, Silva. Do you hear that? I'm not a gentleman, Silva. I'm nobody. Say what you like, Silva, he was a bloody gentleman...'

'He was very sympathetic,' Silva interposed.

Calcott paused, before adding:

'If I had those two young bastards here who let him do it, I'd wring their blasted necks. Side! La-di-da, that's what did it. Damn lot they care. Go on. Hadn't the decency to come back. If I was to say something to the police they'd want to know more about it.'

The sight of the letters on the table made him take them up again and read through the instructions. He ran his finger along the items. He went over to a small wooden box which stood in the corner.

'They don't say anything about this,' he said in a different, quieter voice. He read the letters which were in his hand.

'What is it?'

'Fetch me a hammer,' Calcott said.

The room had originally been Wright's. The case contained three bottles of whisky which had been left behind. When Silva brought the hammer and Calcott, wrenching off the lid of the box, discovered this, he took out a bottle.

'I remember,' said Calcott, riding down Silva's look with a hard, whipping stare. 'He was a damn decent sort. "A little present", he said. And it's been lying here ever since. Here, Silva! I'll drink the old chap's health.'

When they had got glasses he filled them and very gravely handed one to Silva.

'To Charlie Wright,' said Calcott.

Silva unexpectedly made the sign of the Cross.

'Eh?' said Calcott. 'Oh yes.'

'Drink up,' said Calcott, made aggressive by the first bite of the liquor. '"Charlie Wright," I said.'

He drank down his glass, gave a gasp of excitement and sucked his lips.

'That man had a heart,' said Calcott. 'You dagoes don't have men like that, Silva. If it wasn't for chaps like Charlie Wright where'd your country be? You wouldn't know you'd got a country. Take Johnson. He's another. *And* his father.' Calcott at his second glass had warmed to fervour for the men whose necks he had wanted to wring a few minutes earlier. 'They've got hearts. You can say what you like, Silva, a gentleman's a gentleman. If they die they'll die like gentlemen.'

'I liked them,' Silva said. 'But it was too far where they wanted to go. I said to him, "Why do this? You will never find you father."'

'Silva,' said Calcott, struck by this speech and an idea that it prompted, and in some way feeling in his warmth for them that he was helping them if he did this, 'shall we have a turn at the table—see if we get anything? The table'll tell us.'

It had been one of the pains of Silva's absence for Calcott that he had not been able to try the table. He had made his wife try one night but when her heavy hands leaned on it, the table had remained still. Now he begged Silva eagerly and Silva, agreeing to this as he agreed with everything, brought the table out. They closed the shutters, shut out the cries of the children and the shouts from the street. Then they sat down, Calcott's bony hands, with their split nails, spread out like claws on the table, his eyes half lifted to the ceiling, his head on one side listening. Silva, with his eyes lowered, sat still on the edge of his chair, like a modest pianist considering the keys before he flies away into sudden improvisation.

The results of this seance were at first confusing. Calcott said he feared Hamlet was trying to get through and he didn't want him. Not tonight, he said, when they wanted something more important. But presently the unsatisfactory fooleries of the table ceased. A bolder motion began.

'Who are you?'

'Yes,' came the reply.

'What's your name? Here, Silver, ask him his name.'

'Would you be so kind as to tell us your name?'

'Wait a jiffy,' said Calcott, 'I've dropped my pencil. Now!'

But the spirit hesitated. Wounded, he said. How? Gun he said. Pain. Where? On river, he said. How did you get it? Gun went off. Shot. Want help.'

'Funny,' said Calcott. 'Who is it?'

'Maybe it's Wright,' said Silva.

'Good God!' exclaimed Calcott.

'Is your n . . . Ask him what his name is, Silver.'

'Perhaps if you would be so good as to tell us your name we could help you,' Silva said.

'Yes,' the spirit said. Then faded.

'Is it Wright?' asked Calcott.

'Yes, the spirit said. His name was Right. R–I–G–H–T.'

'Something wrong there,' said Calcott. 'Does he mean we've got it right?'

Confusion followed about the name. Silva himself was surprised that Wright's name was spelt with a 'W'.

But once this difficulty was overcome the spirit became voluble and he told them an extraordinary story. He had not shot himself, he had been murdered. 'Hear that, Silver?' Calcott said.

One evening, the spirit said, he had gone out on a river. He and his companion had discovered gold in the bed of a stream. His companion, he said, had seduced his wife, and wanted the gold for himself so that he might be rich and give the woman presents. They quarrelled. Seeing his opportunity the companion had murdered the speaker.

Calcott's arms were dragged mechanically back and forth. He was gaping at the story.

'Ask him who did it.'

A friend, the spirit said. Was the friend there? No. Where was he? Escaped, the spirit said.

The friend's name, the spirit said, was Johnson.

The table stopped. Silva relaxed. It had taken nearly two hours for his fantasy to appear.

'Funny,' said Calcott. 'Open the shutters.'

'What's he mean, he was murdered by Johnson? By young Johnson?'

Silva sat back and considered the work of his fancy with a mingled disinterestedness and admiration.

'Oh, maybe he was telling lies,' he said. 'Like Hamlet.'

'After all,' Calcott meditated, 'we've only Johnson's word.'

'That is what I think,' Silva said.

It is difficulty to say how much Calcott or Silva believed in any of their spirit messages. Sometimes Calcott's readings of the Scriptures warned him that it was the devil's work to attempt to communicate with the dead; when he felt this he said to himself that he did not believe a word. He just liked someone to talk to, he said. It broke the loneliness talking to the dead. At other times he believed fanatically, because he did not know what other explanation to give the phenomena. As for Silva—had Johnson really murdered Wright? Probably not, but the Johnson and the Wright of the spirit messages had ceased from the moment of creation to be the originals on which they were modelled. They had become characters in fiction.

The next day, while he was standing on the ramshackle jetty with the native idlers, Calcott saw a sight which revived the melodramatic question acutely. He saw one of Wright's crew, one of the men who had returned with Silva. He was standing in his boat and he was wearing over his cotton trousers a pair of mosquito boots. The man was very near and Calcott at once was convinced he had seen the boots before.

Calcott went up the wide street, climbing to his house. He was striving to remember where he had seen such a pair of boots before. He was wondering how a half-naked boatman could have come to own them. He remembered when he got to his house and saw the case of whisky. Wright had had such a pair of boots. Calcott remembered coveting them too.

He saw the man again that day. He was lounging against a wall in the street talking to this friends. A theory startled Calcott's suspicious mind in which the remains of the spirit's story still flickered. Wright had been murdered for his boots! It was just the kind of crime the 'dagoes' would commit.

Calcott sat in his room and read again the letters from Johnson and Phillips. Very clearly, they said, that Wright had died accidentally. If this was so, the boots must have been given to the man after his death. Calcott saw Silva and he thought, in spite of the letters—this chap knows more than he tells. Why did he come back alone? He said to Silva:

'Have they sent back everything that belonged to Mr Wright?'

Silva replied that to the best of his knowledge they had. He was a little proud of having had charge of Wright's few possessions, a watch, a compass, a notebook and a box of clothes and books.

'Who did they give his boots to?'

'No one,' Silva said. 'They buried him in his boots.'

'What, they didn't give them away?'

'No.'

'You're a liar, Silver,' said Calcott.

Silva was never disconcerted by Calcott's insults. He regarded them as one of the eccentricities of the English, they even gratified him. A foreign language has always an air of fantasy.

'No, it's true,' said Silva. 'Because the men wanted his boots; but I explained that with the English it would be impossible.'

'Well, they've got them,' Calcott said. 'There's a man in this town wearing them. Come on, you know it. How did they get them? You came back with them.'

Even Calcott, in the drama of his suspicions, saw that Silva's surprise was genuine.

'They must have dug them up,' he said, 'before we left.'

Silva thought this was a disagreeable action, but he could not bring himself to take a horrified view of it. Robbery, murder, bribery, all crimes were not uncommon on the river. As an admirer of the English, he himself took the view that all men were brothers. As a Portuguese, he knew that brotherhood was one thing, property another.

But Calcott was shocked. He was shocked by the thing itself and shocked by the resigned shrug of Silva's shoulders. An Englishman's boots had been taken, a dead Englishman's boots— Calcott rose up with indignation. There was more in this than met the eye. With Silva he searched the town until he found the man and asked him how he came by the boots.

'They were not given to you,' Silva said.

'They were,' the man said. A heated argument and explanation began in the street. The man denied that he had taken them from the grave.

Silva listened to the argument, the protests. His face had the innocence of a young girl asleep, half-smiling at a dream she is making.

The death of Wright, the mysteriousness of the circumstances of the death and the apparent robbery which had accompanied it, acted like an exploding mine under the stronghold of Calcott's Englishness. There was a breach in that stronghold where Wright had been and through the gap streamed the insectile swarm of Calcott's buried fears of the country. He had no defences. He feared more than anything else that he might die away from England. He had seen many Englishmen die. Moreover Calcott had admired Wright's boots.

On this day Calcott's state was one of patriotic property hysteria. It conveyed him to the police. They recalled some doubt about the permits for the Englishmen's guns and ammunition, and the list of the possible crimes of the Englishmen grew in number. Awakening to action, the police arrested Silva and the crew, then paused to see what next they should do. Pausing, they relapsed once more into torpor and sleep and the flies walked on the shining barrels of their guns.

BOOK FOUR

CHAPTER THIRTEEN

IN the first three days Johnson and Phillips were changed. The strangeness, the pain, the sweat and struggle, the boredom and the uncertainty of what the next hour held, the fear of what might lie in the water of the river or be waiting in the trees, the weight of the sun, the blank, dead blue-board of sky so dense in blueness and yet so vacant—all the burden of those days was no longer distributed upon several men with unlimited time before them, but was now loaded upon two. The moment they departed from their companions, they felt this unmistakeable increase of the load. The weight of the light increased upon the eyes, the weight of the green upon the mind, the weight of the heat upon the skin. The country, dispersed before among them, became concentrated, its aspect intensified. Neither Phillips nor Johnson spoke on their first morning alone. When, very early, they passed the creek where he and Wright had gone, Johnson did not look up. He had chosen the opposite bank.

The vision becomes adjusted to smaller fields, life to more exact ends. Phillips wrote in his diary:

'River gone to pot. Spread out into a marsh of puddles and pools and bushes. Sometimes only five feet wide like a sewage ditch or a watercress bed. Awful smells. Everything rotting and the scum drying on the trunks. I paddle and J. stands up in the bows with a pole.'

He was thinking of those old, unused lanes in England which have become an almost impenetrable tangle of nettle, thistle and thorn between the overgrown trees, where the sodden earth smells sour. England going back to jungle after war. You hated war because killing was futile and frightening; yet here you brought the fantasies of your war-time childhood. You dug trenches, built pontoons, took part in imaginary skirmishes. There were many more moments like the one on the launch, when at the sight of the cover of the trees you took that imaginary pot-shot at the mulatto and saw him tumble in the water. Why this trudging, interior stress on the desire to kill?

Immersed in the thicket of the river wilderness, the sky often shut out for long periods when the water became a ditch tunnel, tricked up false channels and barred by some fallen trunk which fell not alone, but dragging a mass of wiry liana and lesser trees with it, like a girder bringing down the roof of a bombed factory—'bombed' was Phillips' word—and so forced to return often upon their tracks they did not think of the surrounding jungle.

'The banks of the river have closed. We are pushing into the hot mess of reed and bush and birds go up crying.' As if there were a violation of the land.

'Sweat on Johnson. Skin goes red rather than brown. Swollen. Mosquitoes get at the inside of your nostrils, your ears and your eyelids.'

When they rested they sat scratching like the Indians.

The bloody fool to be standing up there half-naked, with his red ears sticking out. His shoulder blades worked smoothly under his skin. But it's too red. His skin is too damned red. This red, stocky sweaty man who smells too sweet. Lucy's lover.

His thoughts went over to England again. Lucy's words when they came out of her lips always seemed to stroke the air with the soft summer hum. The hairs above her upper lip were like the fine black hairs of a big bee.

Then he was back at their departure. The limbs of the river-banks were closing.

'My mother used to belong to a very advanced set of people when she was young. I've just been thinking they can't have been as advanced as she thought,' said Johnson out of space from the bows of the boat.

He was in England too. Phillips smiled.

'What is "advanced"?'

A huge, pink winged bird like a heron was flying over the river.

Phillips did not speak as loudly as Johnson. This was partly from fear of being heard in the trees and partly because he felt everything they said would be ridiculous in this place. A cheerful, public school amateur like Johnson could not see that only Indians were not grotesque in this country. Stupid hero.

Supposing one morning, like this morning, they discovered the fate of Johnson's father? Supposing there was a pocket-book, a boot, something metal which lasts, a gun or a watch! What happens? Does he vanish from his invisible leadership? If he is to lead us, if we are to believe in him and if he is to have power over our lives, is it not necessary that he should elude us for ever and be unknown?

Johnson did not often think about Phillips. He was not aware of the body and of the mind of the man behind him, did not know that Phillips' thoughts were so often passing like invisible fingers over his back or settling there like touchless flies. In the evening they had the luck to find rock where there was no mud and the flies were few when they camped.

They put their fire low. It was a delight to cook a speared fish which had caused such laughter and swearing in the spearing. A speared fish, Johnson explained, meant a cartridge saved.

Climbing a high tree before dark they could see no smoke of Indian fires.

'You're all right tonight,' Harry said.

This relieved Phillips, but Johnson was disappointed. Without the help of Indians, whom he intended to approach whatever might come of it, they could hardly leave the river and find their way.

Nevertheless they did leave the river.

Phillips' Diary: 'Fifth day on our own. No Indians. Left boat and now lugging everything. It weighs a ton. Cutting our way and torn to bits. Have gone lame. H. says this is the crucial bit. A sort of porterage without canoe to the Rio Negro—every bloody river seems to be called Rio Negro—reckon it can't be more than fifty miles. Then we're right on the track. I'll be glad to see a river, to let a boat carry this pack. Feel like a Woolworths. Thought snakes and tigers would be the trouble in this country but it's a poisoned toe and an incapacity to digest tough (so-called) partridge.'

Sometimes the country was open, the bush and dry vegetation scratching the sky. It grew out of the mud in the marsh places and water holes like the fibre of a grass broom. Drop a match and the dry scrub would have gone up in flames like hay. The place was glazed and still like a photograph and with hardly more colour. Trees stood up like English trees but casting a dry, savage, uncooling shade; and gaudy fowl, like clockwork in their noise, went out of them. Lizards rippled away at every step, the jewelled and the leaf-coloured, and small deer sometimes in these wrecked brakes, stared with naïve curiosity too long. Once or twice in these days Johnson made a shot so easy that even hungry and anxious men, thinking always miles and days ahead, paused to smile in admiration before they fired and the smile died.

There was a strange joy in feeling that this country was filled and alive and continuously breeding with hot, fearful, hungry animals like themselves, walking in a thin, fine line of ecstacy between life and death, preying upon each other, though-dulled, sun-choked, bruised and torn by the insuring earth.

He did not know how to say it but it seemed to Johnson that no disturbing questions should be asked. But Phillips, ever nervous, restless, his brain and nerves like the taut strings of an instrument responding to every sight and then plunged by exhaustion into melancholy and a quiet maddened boredom, would not leave his thoughts alone. In the silence at midday when there was no sound whatever but the breathing of their tired-out bodies, hidden memories of his childhood, fantasies of his early life came up from the savagery in himself. Johnson listened if Phillips broke their habit of not talking seriously, but said little, revealed nothing. Sometimes he was going to speak; his words rose, simmered and sank back.

What he would have wished to say if it had not been confusion in his mind was that he had passed through the soft horror of Lucy, the hard speechless horror of Wright's death to the elaborate and voluptuous reliefs of living alone and the belief that in discovering his father all was expiated, defined and justified. He did not think these things. They poured like reflections of earthly things over the head and body of one who is swimming under water.

On the march, if it could be called a march and not a slow tottering of two loaded, top-heavy figures with heads rarely looking up to more than a few yards ahead of them, they were together in consultation about landmarks and the compass and when they were held by the undergrowth which was thrown out like the coils of wire of an enemy's entablements. They had to take out their long Brazilian knives and cut their way through. But they walked apart for hours at a stretch, slow hours, slow-moving men, like two ants crawling through grass, moving by instinct from one edge or pool of shade to the next. And when the bush was low and there was no shade for head and body but they must take the huge clasp of the sunlight, hugging them like some fiery golden ape and dragging their heads and shoulders down, they moved into those inches of shade made by tufts of grass, to get their blistered feet into that fugitive coolness. They leaned upright against trees to rest, for it was a labour to get to their feet again under their loads.

Johnson was in front. He turned his head only if he saw difficulty. There was a grin of patience in his beard. They laughed quietly at their ludicrous appearance. They were used to the sight of each other, had no words. Their talk was reduced to single words. And those words soon lost their original meanings and new, ridiculous ones or abbreviations were used. These were more expressive of their situation than the words of ordinary speech, implied under the comedy a trust, a faith in each other, the bond of a shared history. Water was called mud. Farinha was called sawdust and became abbreviated to 'dust.' The tough, stringy game was extravagantly called peacock at first but soon became 'peek,' 'pekinese' and finally 'dog' when being eaten. The compass was 'the jigger.' So 'take a jig' was to take bearings. Climbing a tree to reconnoitre was turned into 'comb the place for lice.' Quinine, though never taken, was called gin. Shooting a bird was 'balling up a hen' and was reduced to 'balling.' Phillips made malicious reference to Johnson's Greenland adventures by saying 'What about balling one of the dogs.' Enquiries about personal health became obscene. Phillips' diary was his 'toilet roll.'

These words, and dozens more, became their only language, covered the essentials of living. They became the language of their trust, their understanding and their bond. Less and less they spoke and the words became shorter; their completest trust was in silence. True, they joked hungrily about restaurants and amused themselves by making imaginary menus; anything more serious they avoided. Normal speech, would have been alien and rich in betrayal. To suggest their normal world would have insinuated doubts, angers and irritations, would have made them separate. When they were together hacking their way, they merely swore.

But the silence between them, that is to say, the faith, was not perfect. In the end there is no unity between people, there is a final, inevitable preservation of the individual. Phillips' faith was more emotional and more vocal. He saw the figure ahead, the drooping back, the broad shoulders, the gleam of the gun-barrel, the dirty white handkerchief concealing the neck, and when he was nearer, the handkerchief blowing back in the occasional breeze and exposing the boils. The bearded face, though he was used to it now, when turned to him, was strange. It was a mask. It was symbolical of the new Johnson, the man of sickness, guilt, obsession, fanatical energy and unexplained decision whose motives were buried, whose existence he had not suspected. With the beard Johnson looked old, cunning, wily, a dirty slogging dwarf on some self-chosen treadmill. His teeth when they showed between the lips looked like the teeth of a dirty gorilla. This was when his face was turned: but most of the day Phillips saw only his back.

Then Phillips' mind was softer and happier, trusting and loving. Yes, loving. All the attraction he had had to Harry in England bloomed in Phillips' heart into a deep love. Sometimes with jealousy he joined with Johnson against Lucy, who had so vehemently robbed him. He saw her again and again on the stairs of Mrs Johnson's house, her eyes brilliant in their dark shadows, her skin too white, her lips too set, calling Harry to her room on his last night in England, and Harry, frowning and hesitant, not wishing to come. She had seemed then in this final moment of her possession of Harry, as she stood in the dim bad light of the landing shadows, like the negress he and Harry had seen when they were coming up the river. Just before Harry was sick.

Then he changed sides and joined with Lucy because he had loved her too and had been joined to her, bathed in her as if he were pressed into the unguent of a lily, and had felt thereafter that they walked into each other's minds like happy people who came smiling into a familiar room and can stay there awhile perfectly at ease, alert with curious interest, and he felt then again what he had felt in the launch at the beginning of the journey, a kind of pity and tolerance of Johnson. He, Phillips, was there to care for Johnson and bring him back to Lucy. In this excited state—and often in the midst of worn-out misery it would come to him, not to stay long, but to flutter across his mind like a relieving breeze—he felt the strength of loving Johnson, a passionate devotion to the man who was trudging ahead. 'He's hard. He's tough. That man never tires.' And Johnson went ahead untroubled, with monotonous unhesitating regular pace, like an ox yoked to his object. Phillips was glad that he and Johnson, alone and sharing the intense burden of this world removed from all other worlds, were bound to the same woman. There came to him in these moments the desire to tell Johnson this.

But these moments, though not rare, began to lose their relevance. The only important question was, would the two of them survive?

CHAPTER FOURTEEN

BUT not to Johnson. He had no doubts about survival. The question was, would they find his father? Had they passed him? Indians would know and could lead them to news to him. Him—what was meant by *him*?

At night they kept a watch. Phillips suggested this and Johnson agreed. He agreed in a laconic, courteous way to a two-hour watch as if he were making a concession to the weakness of Phillips. Through the night the watcher kept up the fire.

'What does he think about when he stands there?' Phillips looked at the figure of Johnson against the flame under the white stars and slept without an answer.

There were many things Johnson could think about; he could argue with guilt, could think of attack, hunger, thirst, death by exhaustion, the impossibility of getting through. He did not think of these things but of the date, the day of the week, the miles done, the correctness of their course, reduction of rations, prospect of water. Only one thing worried him and because of this he continually watched what animals they saw and then the sky, comparing the visibility of the stars on this night with their visibility on another. Every minute of his thinking was occupied with these things, as the second hand of a watch occupies every millimetre of time with its routine. The question that worried him was, would the rains come early this year? Or would they be late? There were freak years when the rains came early, and if ill-luck brought them early the chances of getting through were small in flooded country; sickness and hunger would come with the rains. And here, the Johnson of expeditions, trained in leadership, interrupted the new fanatic man. He looked at night as he sat by the fire engrossed in Pickwick yet aware of any sound, or of any pair of eyes that might be cut like agates out of the dark bush—he looked at Phillips and felt responsibility. Ought they to go back?

He argued it down. They were equal partners, taking equal risks. But the sense of responsibility, lying still in the day, grew in the night when he thought of the rains and the misery of sickness. He stared into the fire. He, like Phillips, had grown thin and his eyes were sunken, intent, hungry in their look and circled by violet rings. There grew in him the desire not to have responsibility. There grew in him the desire to be alone.

In the day following he was always ahead, aiming at increasing the distance between them. Yet he hardly had a full consciousness of what he was doing. Before nightfall, after frying their food in the blackened pan, chewing the tough meat in their teeth and wiping the blood and the fat from their hands on their clothes or the grass, belching like squatting animals, and staring with dyspeptic eyes, they tried to talk. Or Phillips tried. To talk one had to use language beyond the scope of their abbreviations and slang, to relapse into the old buried language, so rich in betrayal. It did not seem to either of the two men that there was a betrayal in this language, there was only a tacit reluctance to use it.

'Until we meet Indians' (Johnson was on his persisting theme), 'we shall not know where he is.'

'He' was the father. By consent the name of 'father' was not used. It is tabu to use the name of the god.

'*Is?*' said Phillips. 'You don't think he is alive?'

It had all been discussed before as a piece of speculation; now it had become oddly personal. Wright had demonstrated in old discussions that the missionary must be dead. A white man could not survive years of life with the Indians, unless he was mad, any more than he could live in a stable.

'He might be alive,' Harry said.

'If he is alive he is insane,' said Gilbert.

'He was a strong man in the prime of life,' Harry said.

Reason argued that he must be dead, but every step was like an act of belief in his life. Little by little the father's life grew, the flickers of wind might be his breath, the brilliance of the sun on the bush might be the light of his eyes. He was there, intangible but unmistakeable, a pervasive presence. Perhaps now he was among those trees, perhaps now in that valley; perhaps now, as they cooked their food, he was cooking his and when they lay down to sleep, he too lay down in some Indian village. Johnson's face became obstinate, the underlip red as a worm pushed out in the beard. What would they say to him when they met him? What would be changed. What would he wear? How would they know him?

Johnson remembered a slim man with large, rather cunning eyes standing on the platform at Waterloo more than seventeen years before. His mother and brothers were there. They, the children, were excited. They were glad that their father was going because it made them proud and because they would be free. They were impatient with their mother because she had tears in her eyes. The missionary took out his newspaper. He was an economical man who bought one newspaper in the morning and preserved it all day. His sons smelled the ink-smell of the newspaper on him when they kissed him. It seemed natural that so much should be written about him in newspapers afterwards.

Johnson remembered other things about him. His shouts to them when he taught them to swim, how he taught them to camp and light fires, how he said Grace before meals, how his voice was sonorous and harsh in churches, how he showed them native weapons, how there was always a quarrel with the missionary society, full of sarcasms on their father's part, about money. Sometimes important people from the society came to the house and then their father would talk in a language they rarely heard there. A very zealous, religious jargon. After these people had gone they often heard their father say with sudden candour, 'Keeping these people quiet is a question of knowing the language.' He always spoke enviously of men who were free of the missionary society. There was also a man called Mr Bunter who lived next door and had been in Malay. He used to come in and attack their father and tell him stories against the missionaries. Their father used then to look powerful, austere and cunning, like a god, when he counter-attacked upon Mr Bunter. But these memories were few and confused and the father they represented was only a fragment of the father of whom Johnson thought. The living father was the man of the last seventeen years or more, the man they knew nothing about, a vacancy rather than a man. He was all the things one did not know. One would feel one had grown up behind his back and that one brought the whole of one's manhood to his judgement. One poured onself into the invisible, intangible, unknown mould of this vacancy.

Phillips looked steadily at Johnson. He wanted to say quietly, 'I want to go back.'

But the question of going on or going back had become academic. Under these stars, in these bush smells, hundreds of miles from help or knowledge, there were two human beings become ciphers, written off merely as missing. Yet wherever Johnson walked there was courage and Phillips drank it.

Johnson loosened their seriousness by smiling. He said:

'He might be king of this country.'

Phillips said, 'The great white chief with an Indian wife, singing psalms in holy bigamy.'

'Me, the returning prodigal, spoiling it.'

'This bloody country's full of your brothers and sisters, I expect.'

They laughed a long time and Phillips prolonged his laugh to live longer in its absurd happiness as if it would be a long time before they would speak or laugh again. He laughed full out.

They were camped within fifty yards of a wall of trees. Daylight had not gone and as Phillips' laughter died there came from the forest another but extraordinary roaring laugh which they thought at first was an echo. It ended abruptly and then— they had got to their feet, looking at the trees—it came again.

The sound was animal, but not the husky abrupt roar of a great cat or any other animal noise they knew or had heard. It was not a very still time for the insects were striking and there were the rattlenotes of known birds, but this sound broke through them all; it was like the shout or gulp of a drunken man, maniacal and sudden and yet concerned with some private domestic matter. It was like the noise of man, but of a gigantic man, the roar of a Lear in madness.

175

The sound did not come again and the evening poured its growing quietness into the wound.

'Charles heard that once,' Johnson said. (This was one of the few times he mentioned Wright's name.) 'He didn't know what it was.'

The sound came from an orang-outang or perhaps a tree had fallen and monkeys had jumped screaming with it.

They sat down and their weak laughter was ended.

'My father,' Harry said, 'did not take anyone with him through this country.'

One took the first watch and the other slept. A joint experience of the rare makes love grow in the heart. Speaking had become this rare thing and when Harry stood up to watch and Gilbert lay down to sleep, each was wide awake with his trust. But as the night went by and sleep did not come to Phillips the trust grew tired and anger grew in his body. He saw the black figure of Johnson sometimes standing, sometimes squatting in the light of the fire, alone. There was no communication. There was love and then the anger of the frustration of love, the hatred of the impossibility of being mixed in that man and being him.

In the beautiful morning, cool as a pond and shining, the two men went on. The mists were like seeding cotton on the ephemeral trees. Every day a shorter journey, every day the heat and weariness, the pains of the body, the greed for food and the need of water coming earlier. They were made hungry and exhausted now by the act of strapping on their packs and of getting to their feet under their grotesque loads of greasy pots and pans.

On the march the body does its mechanical work and the mind prowls around it feeding on the past and the future. The body goes on now, but the mind may be twenty years away in some other landscape, hearing other sounds. Coming upon that place where, the night before, they had talked, Phillips stopped dead in his march. He heard the words of the night before—'My father did not take anyone with him through this country'—and saw another startling meaning in them. The bowed figure ahead of him, loaded under the sun and scarcely visible against the chloride green of the vegetation in the glare, had become a different and more remote person. Phillips realised that Johnson wished he, too, were alone like his father. He wished that Phillips was not with him. Phillips was an irrelevance.

First of all Phillips was in panic. He hurried his pace to catch up Johnson. The pans clanked. In the bush there are deep belts of watching silence, especially as the sun goes higher; then in some places there are belts of sound, from birds chiefly. After that there are long times when the monkeys are screaming perpetually in the tops of the trees, so that the traveller is maddened. In the silences he is awed and alert for the sudden honk, the fall of a fruit, the distant mysterious crash which an animal may have made or which may come from a falling branch; in the cries of monkeys and birds he has a deafened and exasperated longing to be in the silence again. The pans clanked on Phillips' back and hit his thighs, breaking the silence as he exerted himself in panic to catch up Johnson.

Johnson stopped, bending over his compass.

'God, I'm a fool,' Phillips thought. He was ashamed now he had overtaken Harry. But he saw to it that they went on side by side, pretending that his panic was merely folly, stamping it down.

The fear of the country, of every tree and inhuman mile ahead of them, now was streaming in Phillips. When they paused he looked back and he looked ahead. The country behind them was exactly like the country ahead, one view a snapshot copy of the other. Harry wished to be alone! Turning back would be like going on. These were the observations of pure terror and Phillips could feel his scalp stiffen and his hair seem to separate and drag with fright. He was shut in a room with two doors, and both locked. At the height of these panics he saw he was caught. There was no way out.

Then he calmed himself, looking at Johnson.

He knows. He is not tortured like this. He is calm. He observes. He thinks and does not imagine.

All these statements were true except the first. Johnson did not know. All Johnson thought was:

I am responsible. If I were free of this responsibility, if he were not here, I could get on.

This duel went on in silence between them, the distance between them vacant yet electric with this soundless, speechless skirmishing. And Johnson's preoccupations were scientific.

One would be able to test the truth of the saying that a man alone in the bush would go mad.

It seemed to him an immoderate statement. He saw no reason to believe that he would go mad if he were alone.

The dryness of the day became virulent as it advanced. They appeared to be on the long slope of a plateau and the trees went to north, south, east and west in an unbroken circle of green, from which all strength had been taken by the light which burned in it. At noon there was no water.

'No Indians,' Phillips said, 'because no mud. They know it's drying up and cleared out.'

This was the first time they had been waterless. They looked with craving at the blue sky, stepping out of the shade to search it fully. Johnson had said casually, 'With any luck we shall find water.'

A total change of mood had come upon Phillips. He had not been light-hearted or serene in this march but there had been something romantic in it for him. He knew that they were taking great risks and that if some audience could watch them they would be considered gallant. All day he created imaginary audiences. It would be something to boast about. The chief glory of their folly went to Harry, who had gone imaginatively and with his dogged will upon the project; but glory was reflected upon Phillips.

Phillips in these days was seeing a dream of courage become real as he followed Harry. The dream was the real motive of his departure from England: to prove that he could master the private current of fear that was always drifting against him in every moment of the day and diverting him with imperceptible dissuasions from the course of his imagination and will.

Following Harry he found an increase of this mastery. Virtue came out of Harry as he led and poured into Phillips as he followed. The miles ahead, when they stood level with each other, paralysed the will, but as Harry moved off into them the very earth and its trees seemed to lose hostility, every yard taken became safe. And through the days a struggle not to follow Harry but to be level with him took place in Phillips. He attempted to gain on him; and miraculously, as every day died, fear diminished until he felt on the brink of Harry's fearlessness. Seeking the father, he was to be redeemed by the son.

And then while he was on the very brink of this redemption, the door had closed on him. You could see this. Courtesy and affability drop away in camp. There was now almost no concealment on Johnson's part. He wanted, quite bluntly, to be alone. The fact preyed on Phillips' mind. Harry wished to be alone, thought it simple, easy, desirable to be alone in this place where even being with a companion was fearful and dangerous. Rejected by the son! Denied a place beside the hero! Contemptuously treated, unredeemed, irrelevant!

When men are living in their communities among their fellows, they can forget about these things or work them off in a dozen ways. A few solitaries become morbid, create works of art, are ill, commit the occasional crime. In sick societies they take political power and become tyrants. But these two men were alone. The joy of realising a dream changed into the disillusion of seeing it was a dream only; he could not have Harry's courage. In that unguarded excess of admiration for Harry, Phillips had had, there had been, as in all love, a germ of hatred, which the fanaticism of solitude could multiply into a nucleus.

'He wants to be alone—then let him be alone. Let him die.' Let him die means, 'I will kill him. I will show who it is that chooses to be alone!'

'He wants to be alone'—that meant humiliation.

'I hate him for humiliating me. I will avenge myself.'

But in all crises where Harry was concerned there was this thought to fall back on: 'I am with this man for Lucy's sake and I am more deeply Lucy's because I am not her lover any longer, because we passed through love and are in the kind of immortality of having loved. But he and she are muddled together in the temporal struggle of love, wounding each other, kicking and bruising like animals on heat, half angry with the force that has possessed them against their wills.'

So his mood rambled on: 'He humiliates me. All right. But he can't. He can't be alone. He can't be alone from Lucy or me because we are all bound up in this. We are joined through Lucy. It is inescapable.'

He saw Harry stop to read the compass. 'He has got to know this.'

Every stumble and jerk of the marsh, every stab of pain in his limping foot, increased Phillips' determination.

'I suffer now but for every stab you shall suffer. This is a joint pain. You deserted me.'

It occurred to him that the way to hold Harry was to tell him that he (Gilbert) had been Lucy's lover. That would arrest and hold him, make him see the bond.

He sat by the fire preparing to tell Johnson that he had been Lucy's lover. His hatred was swelling. All Phillips knew was that this man had a damn-fool, cocksure genius for trying to kill himself; the muddler with women, the red-faced average man of moderate opinions and bloody silences because he had nothing in his head, had landed them in this camp without water. He began in malice, to whip and bludgeon this clod into seeing he wasn't infallible, couldn't be alone, was no better than anyone else.

But when he opened his mouth to speak all he could say was:

'I suppose you know we're going in a circle.'

'Oh, how's that?'

'Of course we're going in circles. You always do in the bush. You must know that.'

Johnson was surprised at the taunting tone, but said calmly:

'In that case we'll be back at the river. We'll get water.'

'We're not near the river. We're lost. You know we're lost. I said turn back two days ago.'

'I know what I'm doing,' Johnson said.

'Show me the map then, go on, show me,' Phillips said. 'If you know.'

Johnson began to draw out the useless map.

'No, put it away,' Phillips said. 'That map's no good. We both know that.'

'You can go back,' said Johnson. 'We've marked the trail. I'm not going back.'

Phillips crawled away. He dreaded that he would break into tears before Johnson. He came back after a long absence and when he had sat down he asked in a calmer voice:

'We ought to know where we stand.'

'I'm going to the river.'

'And if you can't get there?'

'I can get there.'

'Well, I can't,' said Phillips.

'We can go our ways. You go back. I go on,' Johnson said.

There it was, said openly. He wanted to be alone.

Thirst was round their throats squeezing like a closing hand. A throbbing was at the roots of their tongues, rhythmic and to each distinct, and quietly it spoke under all their thoughts and words, like the bubbles of a spring breaking into the bed of a pool, saying, 'Water. Water. Water.' As yet it was only the throbbing, consistent but quiet. Against this the counter-assertions were made: water tomorrow. Water tomorrow. Dew in the morning. Lick the dew in the morning. Awake early before the sun blisters it off.

No quiet speculation by the dry, splitting flames, but assertion against the companion: I've got my pain. Why do you obtrude yours? Can't carry yours. A bound man can't undo another's bonds.

'You go back. I go on?' Phillips repeated, struggling with his horror.

'Yes.'

Here was the limit of Phillips' endurance, the freezing summit of his fear.

'Oh no,' he said quietly.

'Why not?'

I shan't let you,' Phillips said.

'You're not going to do it twice,' Phillips said.

'What twice?'

'Like you did with Wright.'

Johnson turned speechlessly to Phillips as if he were going to strike him. His first was clenched. He said quietly:

'That was dirty, Gilbert.'

'So it may be,' said Phillips, with what vehemence his torn throat allowed. 'But I didn't come up here to die. You did desert Wright. You wanted to on the river. I don't say it was your fault he died but, remember, some people will. Some people do already...'

'Who?' said Johnson quietly as Phillips' manner became more bitter and excited.

'Silva for one,' said Phillips, getting breath. 'You told a lot of things to Silva.' Phillips' jealousy came to its head. 'Well, Silva told a lot of things to me and he'll be down at Calcott's now telling Calcott.'

Now Johnson was stirred. The old dread of scandal returned.

'What's he saying?'

'Just that you murdered Charles because his wife was going to have your child.'

'Good God!' Johnson said and gaped at Phillips.

'You don't believe that?' Johnson said.

'No,' said Phillips. 'Of course I don't.'

'Silva's saying that!' said Johnson.

'He's saying it all right.'

Phillips' outburst had taken him to things he had not thought of saying. The buried resentments came up, the hidden jealousies. His tortured voice was sneering and venomous.

Johnson listened in amazement both to the news Phillips gave him and the bitterness with which it was delivered. But if Phillips had started with the belief that what he said would, in some mysterious way of the will, make Johnson return or would punish him for wishing to be alone, he was wrong. Johnson hated rows and they always occurred in expeditions. The only thing was to be alone. Quite alone, in this country where his father had been alone.

Phillips lay back exhausted by his outburst. He lay back and hid his face from the light of the fire, thrust it into the darkness. He was ashamed.

He was ashamed because he had shown that he was afraid. He was ashamed because he knew he would not have the courage to go back alone and because he was so dependent upon Harry. To be level with Harry and his equal—this yearning of their journey together— was shown to be impossible. He would be cut off not only from Harry but from courage.

But there was a way in which he was the equal of Harry. There *was* a thing which made them equals. It was a secret way but now it must be openly and desperately avowed. He was not the equal of Harry in courage, but he was the equal in love. Lucy's voice came to him as he lay there, begging him not to speak, repeating her words, 'I'm afraid of hurting him. Look after him for me,' but Phillips drove them back. Harry had got to be hurt. Every stab of pain must be paid for. He must be brought to his senses by the shock of pain and since physical pain seemed to leave him untouched, it must be pain in the soul. He must be shown that they were not separate but one. He must be made to know the bond.

The malice, aggression and other motives of the frankly selfish kind which may have been in Gilbert's determination were married to more creditable convictions. He knew something of Harry's love affair and he knew that Harry's behaviour in it had had, under his reluctance, some element of the perverse. Harry had made Lucy suffer, he had made Wright suffer, he was now making himself (Gilbert) suffer. He was cheerfully distributing among others, without knowing it, some surplus of his own self-torture. This would have seemed morally indefensible to Gilbert in any case; but now, since he loved Harry, his heart was passionately concerned for him. The weakness of Phillips was that he thought it sufficient to show people the self-destructiveness of their ways by awakening their reason and that this would cure them. His absurdity and his intolerance sprang from the delusion that he was disinterested. In his egoism he completely ignored the instincts of people, the predatory nucleus which is the basis of human life. He really wished to annihilate with his reason, though he would have been horrified to hear that this was so. The truth he did not see was that he wished to save Harry because he had become the rival of Lucy for possession of him.

Now that he had decided to speak there remained only the difficulty of speaking. The stamping heart had to be quietened, the trembling lips to be steadied. The blood throbbed in his ears when at last he said, gravely:

'Harry, I want to speak to you about Lucy.'

Harry did not answer. He was lying propped on his pack looking at the fire.

'We've got a duty to her, Harry,' Gilbert said. He stumbled on since Harry still did not reply: 'Not only you, but I have. Both of us. I think that is important. I think it's more important than finding out about your father.'

Harry turned his head towards Gilbert.

'I'm not going back,' Harry said stubbornly.

'It isn't that,' Gilbert said. 'You're not going back now. All right. But you're going back next leave. Or she'll be coming out here.'

'Here?' said Harry.

'Yes, here.'

'I'm not going back,' Harry said.

Their voices weakened, strained by thirst. 'I'm *never* going back.'

He turned to Phillips and repeated:

'My father went here. He stayed here. He lived here for seventeen years.'

Phillips sat up and stared in horror.

'You're mad, Harry,' he said. 'He's not here. You know that. And you know you'll never find him. And that you can't stay here.'

The madness of the intention so horrified that it left Phillips incredulous. He fought it off, brushed it aside.

'Lucy loves you,' Gilbert said. 'You can't treat her like that. She loved me and I loved her. I still do love her. I don't mean it is the same as you and she.'

In a stone-dead voice Harry said:

'It's all over now. I was alone before and I was happy. Suppose,' he suddenly turned to Gilbert, 'suppose she's having a child?'

'She's not having a child.'

'How do you know?' said Johnson. 'She may be having one.'

'That's what Silva's saying,' Phillips said. 'Did you tell him that?'

Harry got up without answering and began putting sticks on the fire. Phillip's fear of a high fire at night which could bring the Indians distracted him.

'Leave it alone,' he pleaded. Did Johnson *want* to have them both killed?

'There's no need to be afraid of Indians,' Harry said. He sat down again at last.

'You think she's going to have a child and yet you stay out here. You just leave her to it?'

But, of course, there was no answer Johnson could give to such questions. He lived—didn't Gilbert understand?—in the shadow of the death of Wright, a greater crime had swallowed up the lesser. The first crime he knew with his reason to have been a fantasy, it was part of that unexplained litter of experience that one drags about after one. Some day one will see it clearly; some day one will understand it. Not yet, but some day. But the death of Wright was no fantasy. He had seen the man die and he knew why he had died, and he thought he read on his face in death the inscrutable expression of the betrayed man. And there was also the frightening, thrilling, overwhelming knowledge that because this man was dead you were set free. And the inevitable reaction to that freedom was to choose some path of expiation and make freedom arduous. Johnson answered honestly:

'I don't think she is having a child now, but I thought she did not write me because she was going to have one. I was worried by that.'

'I would have been proud to have Lucy's child,' Phillips said.

'She is lovely and warm and she's got a lot of sense. There is a touch of something savage in her—like you, Harry'—Phillips smiled—'but she's clever enough to hide it. You only know a woman when you sleep with her.'

It was for Harry now to look out from the shadow of the death of Wright upon his life with Lucy and with Gilbert; and doing so, he recognised how much he and Lucy had depended upon Gilbert's goodwill, and how, at first, when he was repelled by Lucy he had often thought: 'I am a fool, because Gilbert likes her and she likes Gilbert.' It was the capacity for quiet intimate friendship among people which Harry most respected. He loved the sanity of friendship even while, alone, he was drawn by the element of insanity and obsession in his passions.

It was Lucy, sane and generaous, who had shown him his insanity and had set about dissolving it in herself; and Gilbert confirmed what her instincts essayed.

And yet, under the shadow of the death of Wright, something in Johnson rose fanatically to keep this view of the past at a distance. He spoke now as one who is describing experiences which are mere reminiscence and beyond the reach of present action. He was committed to the necessity of finding the father.

Harry said without jealousy, without any espionage of mind and heart, but simply:

'When I came back to England I thought you and Lucy were in love. I had never thought of loving Lucy because I'd known her so long. She said she was fond of you. She said once she was always half in love with you. I said to her once, "Have you ever had an affair with Gilbert? It would never surprise me." And it wouldn't have done. I couldn't quite believe it when she said "No." You seemed to understand each other.'

It may have been Lucy's denial, the affront of that; it may have been that jealousy which seems to bring love to the brink so that it may swell over and declare itself. But the occasion, at least, was clear; it was to show Harry that he could not be alone, to keep him there at his side in the bush where the trees were made of fear and every sound plucked with cruel and delicate fingers at the heart, to kill the implications of the words 'Never come back,' that Gilbert said:

'I *have* slept with Lucy. I've always wanted to tell you. Of course it was before you and there as nothing between us, nothing like you and she. We understood each other too well.'

An effusion to tenderness was in his face, a boastful candour. He turned eagerly to Johnson with beating heart, ringing with happiness, and underneath: 'He cannot be alone. He cannot be alone.'

But Johnson's face was startled. His mouth had opened. His eyes were small and staring. A small stick was in his hand. He was banging it on the ground, banging and banging, till it snapped in two. Then he threw one piece after the other into the fire and watched them burn. Phillips was checked. He felt suddenly foolish, abysmally silly. And mean. The stupidity and meanness! His soul shrivelled.

'I thought she must have told you,' he lied.

He went on hurriedly to explain about Lucy and himself, in a panic to heal his own wound, horrified, cursing himself for having spoken. Johnson listened without speech. He picked up sticks beside him, snapped them and threw them into the fire, reaching out for more.

'She told me a lie,' he said, in a voice of unsurpassed bitterness and looked, stiff-lipped, small-mouthed at Phillips, looking at his head, then his neck, his chest and slowly down his stretched-out body.

Phillips was lost and helpless. He saw Johnson get up and—odd gesture in their filthy state—brush the earth from his clothes. He went out into the ring of jumping firelight and, pulling out his knife, cut at the thicket. For half an hour he worked thereabouts. There was the sound of his knife on the branches, the drag of the broken ones as he pulled them out. A mania for collecting fuel possessed him. He slashed and cut with method and energy, absorbed in his task. The heap grew higher and further afield he went and dragged in more. Phillips saw his working arms, his bending body, the fire shine on his knife and on his sweating face. At last he began to bear his load nearer to the fire and Phillips. Startled out of his horror at Johnson's alienation and distance, the sudden aloneness which had come upon him, making them more than ever apart instead of uniting them as Phillips had absurdly hoped, he rose up to help. They laboured silently at the pile and speech dried in Phillips' throat; he waited for a word to release him. But not until all was carried and Johnson had heaved a great pile on to the fire, which crackled and sent up a soft speckled brush of sparks into the sky, did Johnson speak.

'That'll bring them,' he said.

The expected release was choked back by these words. Phillips looked appalled at the flame and its flight of sparks. It was feet higher than the invisible line of safety he had drawn. He lay down under his beating heart and Johnson took that watch. Phillips watched him and the fire, listening to the noises of the bush, and time passed. It passed and slowly his horror sank into him and became part of his life, new wheels of anxiety to work in with the other familiar wheels. His horror was assimilated. Another burden. He grew used to it. And now there was renewed in him the pain which the conversation had silenced for a time: the tightening hand round the throat, the sore throbbing at the root of the tongue, the underlying speech of the body: 'Water. Water. Give me water. Rise very early and lick the dew from the leaves.'

In the morning they travelled miles from each other. Near at night, they were already like ships, still in view of each other, the wisp of smoke on the horizon, but going wider and wider apart.

The day before they had drunk poor water, a thick muddy liquid which coated the inside of the mouth and the teeth and increased their thirst. At their last camp they had camped high and there had been none. They had half-emptied their bottles in the morning when they had made coffee. Now their mouths were dry with thirst. They were silenced. Johnson frowned at the bush.

In the afternoon their way lay under the shade of trees. How long can a man live without water? They licked their drying lips, sucked stones. There was little fruit in the trees.

Their thirst of the morning became a fire burning in the throat and closing it. The shade so cool and silent which a breeze freshened, the ground gentle, the trees full of sap—and yet no water. Their throats were choking. Johnson's eyes grew wide and staring under his frown.

Of course they had known that this might happen. Every day they had said, 'Lucky.' Yet—in the midst of trees and vegetation, with other animals living among them—thirst had seemed impossible. Now it came to them suddenly, gearing them to an inexplicable tyranny. From their ribs to their necks they were caught in its passion. They stopped often to listen for the sound of water; like thirsting animals they began to smell the air. They looked to the earth for spoor leading to some spring.

The marched until the day closed and camped clear of the trees. Soon the flame went up like splitting spires at the dry sky. At night when they lit their fire they hoped and feared that Indians would see it. Johnson kept the fire high in his watch but Phillips let it die. His nerves were on the edge as the flame leapt higher; he came to measure a height of flame which he thought was safe. He winced and bit his lips when Johnson raked in a pile of broken stick and re-charged the fire.

'It's all right,' Phillips said. This persistent quarrel about the fire!

This was their second waterless camp and the fact silenced them, arrested all their thoughts, shocked them.

How had it come about that there was no water? Was it a mile away, ten miles away? Where was it? They looked beyond their camp. The sky and the land had suddenly become a wall against them, a glazed impervious surface.

They averted their faces from each other, ate in silence, stood up, went about in the ring of the fire's glow alone. If they happened to approach each other they halted, made a pretence of picking up wood and passed without a word.

A small dark cloud went over the sky and they looked up at it, each unaware of the other.

'Cocktail,' said Phillips.

This, they both thought in different ways, is the beginning of it. Of what? Of saving the white thing called your body.

But as the afternoon closed and the calls of the evening insects shrilled and grated like cheap tin instruments, and flies came out, and the blisters and thorn tears began to sting, the river floated out of their minds. A well, a spring, a mud hole was now urgent to them. They spoke of it. Phillips insisted that they stop and explore for water alone. His mind had become narrowed to this point of purpose. It had become the only thing.

And Johnson obeyed him.

The whole condition of their journey had changed. Route, objective, Indians, alertness to danger, even the river went. Johnson no longer led, Phillips no longer followed, there was no father. There was no leader because gradually there had been borne in upon them the futility of going anywhere, or of being anything. Thirst was the leader.

'The trees grow,' said Johnson quietly.

Phillips looked at the trees which grew while he and Johnson burned and weakened. Their mouths were like charred holes.

Phillips took out his bottle and held it over his tongue. He tried to imagine a single drop falling upon it.

That night they sat awake by the fire and the roots of their tongues grew, like a hand wrenching down their throats. Many times they got up to feel the leaves of the bushes for dew and licked them, imagining dew was there. Only their physical weariness gave them two or three hours' relief in sleep, and they woke before dawn shivering and astonished; for both had dreamed of the morning mist and the dew cupped in the leaves and water curling before them.

They looked in stupor at each other. There was an irony in their eyes. It would be easier to lie down here in the shade and watch madness creep upon them, rather than to go on and seek it.

They did not eat but at once got up and open-mouthed went on.

'I can't go on. Why does he go on ahead? He's always ahead. I can't go on. God, I can't shout to him.'

Phillips staggered with the blood deafening his ears. Johnson turned. His mouth too gaped open. His eyes stared and glared distended in his thin face. His teeth were pronounced as they are in a skull. They gave him the air of grinning diabolically. Phillips sank faintly to the ground. Johnson did not move. He stared a long time at Phillips, then he too, forty yards away, sat down and waited for Phillips to rise. A last, since Phillips did not rise, Johnson went back to him.

Phillips said, 'I'm finished.'

'Lie down,' Johnson said.

They sat side by side muttering, looking with pity at each other, the perfunctory pity one might feel for a dead rat. The flies came on their lips. Sometimes the breeze, more active this day in mocking freshness, seemed like the sound of water.

Johnson unhitched the pans and the sack of meal that were tied to Phillips and took the extra load himself. If there was space in the heat to contain feeling about each other, the feeling was hatred. A whine came into Phillips' voice, a lameness in his stagger—the lameness of his bad foot—his mouth gaped, all hateful things. The straining of the neck from the loaded shoulders, the blank, glazed stare of the eyes, were hateful. A peremptory irritable terseness came into Johnson's voice. Then under his beard he looked like some seedy Christ. Hateful also were the dressings on his boils. The life of each, the separate existence of each with its tacit claim to a share of help from Heaven, was hateful.

They chewed leaves but these were bitter and stinging to the tongue, the green mash dried on it and they had no spittle, only a film of white foam, bitter stuff from the foul stomach.

'Christ!' muttered Phillips continually. 'I must drink. I am terribly, terribly sorry to have to state your Royal Highness, father and mother of all, I must get some water.'

If he doesn't find that bloody river I'll kill him. I'll pot him off like I did that chap on the launch. Oh God, what a fool I was to come with him! I remember going from the river. It dropped away behind the rock. This is how it would end. I knew this. All my life I have seen how everything would end. Lucy, your two bloody lovers.

And Johnson: Without water one can last seven days. I'll leave him and bring back water from the river. A mile is taking us nearly an hour. I can manage a day more. I'll leave everything except gun and pan and go light. I'll mark trees with an axe. I'll get to that water. If one of us goes mad he must be tied down in the shade and all arms must be taken away.

Phillips stopped. Through the veil of lesser trees, bush-bearded and wired in by thicket, was a small cliff of grey rock with the northern side deep in shade. There had been rock cropping out in the land for days but this cliff-faced pile, thirty feet high, was like the ruins of a castle. He stopped. Fontainebleau, he remembered, rocks like these, a girl with ringleted hair, then photographs of buried Maya temples.

'Harry!' Didn't hear.

He waited. A smile—he thought it was a smile—came on his lips as the shadows of the trees passed like wands over Johnson until he was not a man but a ghost in the trees. Gilbert smiled to think that this was the last of the fool, going further and further away, leaving him behind. Resignedly: 'He'll be alone and I in Fontainebleau listening to nightingales.' Watching him go, the sunlit man, the shadowed man, the sunlit man with a gun, the shadowed man: 'He's alone. He's alone. It's the last I saw of him Lucy. He was going on.'

Phillips stood there. A vulture was on the rock.

'You shall not be alone,' Phillips called out with all his strength. The hardly audible cry made him fall to his knees as if he were in prayer and at this moment Johnson turned and again saw him kneeling. This time the two men did not move for a long time. They merely stared at each other.

CHAPTER FIFTEEN

IT amounted to this now: there was Johnson's instinct that salvation lay in going on to the river, that every blistered step took them nearer. To stop was to cease to act. What did one do if one ceased to act? Souls made to act fester in inaction.

There was Phillips' instinct that salvation lay in stopping now, here, anywhere, but in stopping. In action there was nothing but futility. One must resign oneself a thousand times a day, accepting and being still. One must die in advance. Man dies from the moment he is born. Wise men are inoculated and prepared by a myriad small deaths.

So each saw in the other in this extremity the enemy who wished to destroy him. Each saw the opposite of himself. There were looks of ironical pity in their faces, a dumb questioning, a bond almost of love and then a wild, alert apartness from each other.

To both of them it was mysterious and intolerable that the creature who was bringing him the threat of death was alive, passive, suffering a shared pain and moving before his eyes. Each began to watch with fascination the movements of the other's body. They lowered their eyes. They looked at each other's hands; the hairier, slender, strong and nervous hands of Phillips whose strength was in his wrists; the blunt, short, thick red hands of Johnson with wide-set hairless pores.

Johnson had consented to go to the rock. Leading the way he had cut slowly, his great strength diminished, at the thick mesh of scrub, and the birds flew up, flying low between the trunks of the trees.

Under the shadowed wall where they sat, the rock had split into a tilted platform two yards wide and the top was visible. A stunted tree grew there and one branch of it was the habitual, whitened perch of the vultures. The bird they had first seen was now joined by another and they perched there, rising from time to time to clap their moulting wings.

The nightingales at Fontainebleau sang all night in the moonlight. He must love his father.

My father is the flowing river where I could drink. He went alone without an Indian.

Johnson unstrapped some of his load. He had endured great pain from his boils, but the pain borne in silence had concentrated and made narrow and keener the direction of his will. He put down his gun and unstrapped the bandolier of ammunition. Once they had laughed at this.

They watched each other. Their movements were intolerable and mysterious to each other. It seemed to Phillips that if they could move away the boulders of the cliff one by one they would come to water. There was water everywhere under the earth where the trees drank into their roots. Johnson did not know that this was the reason Phillips lay on his belly and scratched fitfully at the surface of the rock with his fingers. Phillips did not know that, in their circumstances, even this rest was surrender to Johnson and that was why he got up soon to grope about the rock.

The quick dry shadows of the vultures swept over the hay-coloured bush when they left their fouled perch on the tree. Johnson had been watching the birds.

Presently Phillips saw him load his gun. There was a report and Phillips' eyes widened with wonder at the sight of Johnson firing at random at one of the birds as it swooped above; and he was amazed too, even in the remoteness of his agony, which dulled the sound of the shot to his ears, to see that Johnson missed. They were short, too, of ammunition. Johnson stood half-grinning but when he caught Phillips' eye on him he stiffened. Phillips could see he was trembling. Angrily, proudly, Johnson went away round the corner of the rock out of sight.

When after some long time, Johnson did not return, Phillips felt the screw of panic give an extra turn. Steadying himself on the rock wall he followed by the way Johnson had taken. The rock was no more than fifty yards in circumference and he had gone half this distance, too weak to call Johnson's name. There was no sign of him. Then Phillips saw the gun lying on the earth. He stepped forward and picked it up quickly with a spark of cunning and delight. He had always felt some jealousy of Johnson's ownership of the gun, and of the way he assumed that he was the warrior and the hunter of the party. It gave Phillips a peculiar sense of triumph to have found the gun lying there; and at the same time, since Johnson cared for it so constantly, saying that if it failed they were done for, to find it lying there was disturbing. A dropped gun! Is he lost? Does he wish to surrender? One should go unarmed to the Indians. So disturbing that now Phillips called. There was no answer and he called again, weak in voice, waiting for an answer than for some nameless response from the bush. If a thousand heads had appeared in the trees in reply he would not have been surprised. He called again, looking round him. Up in the sky the birds, startled by Johnson's shot, were still wheeling very high, like two bits of charred paper tossed up by the draught of a fire.

'Harry!'

His dulled ears were a sea of dizzy sounds. Now if Wright had appeared, or Silva or Calcott, or if Lucy and Mrs Wright had come there quietly talking about the house or their car or some ordinary things of the life in England, and had turned their heads as they passed, hearing the call, Phillips would not have been surprised.

Then he had enough clarity of mind to be startled by the fact that he could have credited such a possibility, and as these people receded from his mind leaving it bare and clear and firm, he understood as he had never done before the absolute isolation of their position, that in the bush there was no one but themselves. For scores of miles no one. It was revelation and vision to him. For the first time, in these months, he felt absolutely free from fear. He had the sensation, even in his weakness, perhaps because of it, of complete liberty, as though, free of any past life, he had alighted in this place out of the sky, immaculate.

When the spirit experiences such moments it trembles like a liquid brimming in a glass. Not to lose it, he found his way back to the platform, to dwell in this exaltation in the protection of the place he knew. Johnson was standing there. The muscles of his face were moving as if he were tightening it with immense self-control.

'I've been yelling,' Phillips said.

'There's a cave,' Johnson said.

Then he added, frowning as if forcing all his mind into the words:

'Wright is there.'

And after he said this he hung his head.

Phillips' blood went cold.

'Sit down,' he said. 'Shade.'

'In the cave,' Johnson said. 'Come along.'

'Just a moment. Sit down a bit. Fagged,' Phillips said.

'Wright,' Johnson said in a monotonous voice.

'I found your gun,' said Phillips to distract him, 'I found it over there.'

This had the effect he scarcely hoped for. Johnson's expression changed at once.

'God,' he said, with a sudden sigh. 'Give it me.'

'Sit down then. Rest.'

Johnson had had more acutely than he the same delusion. Again surprising Phillips, Johnson obeyed and sat down.

I'm not going to give him the gun while he's like this.

A look of extraordinary hatred came into Johnson's face, a look of sickness and disgust and cunning, as if he had read this thought.

They sat there until they had calmed and had become conscious again, after this blessed interval of hallucination, only of their physical pain. And with this, new suspicions returned. Phillips saw danger now only in Johnson, resented, hated him for his sickness of the mind, as a bird attacks the sick bird: Johnson suspected Phillips because of the possession of the gun. He had forgotten already in the space of a few seconds that he had talked about his hallucination. The two men watched each other. They suspected each other of madness.

He's got the gun. He might kill me.

If I let him have the gun he may turn on me.

The gun was lying across Phillips' knees. Johnson, pretending indifference, saw him discover that it was loaded, saw Phillips—as it seemed to him—playing with the trigger. Suddenly, exhausted, Phillips dropped the gun off his knees and, crawling a few yards to the edge of the rock, vomited. The sourness stung his mouth and he lay flat and spent. Johnson sat staring at him. They had not the strength or the will to help each other. Their friendship was remote in their minds, some distant country from which they had come. The bond was now animal.

When Phillips got up and staggered to the place where he had been sitting he saw Johnson. He was sitting still and upright like a man muffled in fever. His chin was raised and he had a look almost of exaltation in his pain. It lifted his chin and the gaze of his eyes was directed upwards. His hair was shaking in the slight wind. The barrel of the gun shone in his lap. He had picked it up.

No weapon. Can't defend myself. To be so exalted and alone, shut up in bloody self. His father! The stupidity of it. My stupidity to let him bring me to this. But I've seen him afraid. Oh yes, he's been afraid. The last night in England, terrified of Lucy and Lucy telling me about him. She and I were equals. She and I were not afraid. We did not wish to fly to savagery and solitude.

Johnson's eyes were still raised to the sky. They were raised to vacant heaven.

'Water. I must have it now. Water. Harry, water. Damn water. Give me water, damn you. You brought us here. Give me water or I'll kill you. I'll go mad. Are you mad? Get us some water!' Phillips screamed suddenly with all the force of his lungs, which seemed to burst like bags with the effort of his shout and to send gulps of blood pumping into his ears and to blacken his sight. He screamed and the country broke into thick green, blood-warm waves towards him like a sea that has suddenly swelled up to fall upon the sight and the head and shoulders.

Johnson did not move or turn his head. He had not heard. For Phillips' scream was nothing but the murmur of man passing into delirium.

When at last Johnson did turn his head and see Phillips lying face downwards on the rock he considered him for a while and understood. There was a pencil in his pocket. He took out the *Pickwick* and wrote in the inside of the cover:

'Gone for water.'

He strapped on his pack and then considered the gun. At last he pushed it over the rock to Phillips' body and laid the bandolier beside it. Quietly, without looking at his friend again, he climbed down the rock and went out by the slashed branches to the way he had taken before. There were three hours of daylight.

CHAPTER SIXTEEN

HE was on the rock. He was lying with his face to the ground in the shade. Harry was saying something to him but there was only one word he understood, their only word: water. It gasped in his head. Water. Water. Water. Like a dog panting hotly in his ears until it flowed and woke him up. He opened his eyes and saw that in those few unguarded minutes he had been deserted. He was alone. His hand went to the gun at once.

He sat up and saw the desertion happening under his eyes— Harry was off the rock, he was through the bush, he was among the low scrub, making in the direction of the sun. It was on his head and shoulders. He looked like a small, two-legged golden animal in the scrub going lightly forward.

'I am off my guard for a minute. I am on the point of death and he deserts me. He has the strength. He goes off like that slyly, just as he did from Charles. Charles was nothing to him, Lucy is nothing, I am nothing. It is what has been in his head all the time: to be alone. He is mad.'

With the gun in his hand Phillips got to his knees. He tried to shout and no sound came. His open mouth was stiff and made only a fixed hysterical smile. He looked at the rock and saw the note pencilled on the paper of the book, 'Have gone for water.' He utterly disbelieved that Harry had gone for water. Or, if he believed, it could not alter the fact that here he was alone, absolutely alone. He was alone, like the missionary, Johnson's father, had been. The Indians had deserted the missionary and he was alone and that was the end of him.

Phillips saw his fear made flesh and he had no control of himself.

'Come back!' he called frantically, but again no sound came. There was only the stiff grimace of shouting on his face, the unfinished smile.

There was that degree of terror one reaches at once after

waking too suddenly and which becomes a kind of spiritless wonder. There was hatred in it also. The smile was the sign. It was the smile of a dying man who sees his life ebbing away before his eyes. Every step of Harry Johnson seemed to Phillips a ticking away of the final minutes of his life.

'Harry!'—a soundless call, this time finally soundless, without even the sound of breath.

They had both smiled in the way Phillips smiled now when Harry raised the gun to some unwitting animal, smiling at its innocence and the delicacy of that suspension of all creatures between life and death, so easily and chancily broken. There was a moment of love and wonder before one's smile went and one clicked and raised gun and fired. Now, before Phillips' arrested eyes, Harry was going forward through glades of sunshine and shade, like one of those alien and innocent creatures. The blood rose to Phillips' maddened head. He raised the gun and, shoulder trembling, hand shaking, he took aim at the figure of Harry Johnson. He fired. After the shot he did not lower the gun at once but stood considering. It was marvellous to see the man stop in his stride.

On the launch a man buckled and went over the boat side into the water. For a few seconds there was the same ecstasy of horror. Phillips lowered the gun and leaned forward with his mouth open in hunger for the blood. Then he trembled with panic; for Johnson did not fall but turned round and looked in Phillips' direction and in time to see the gun raised at him.

Their faces were not clearly visible to each other. Neither knew what was passing in the other's heart.

Harry took a step or two towards Phillips and called. The sound was not clear. Phillips did not answer the call and did not signal. Awakened by the shame of the failure of his shot, He lowered his head and walked away to the shadow of the rock out of Johnson's sight.

The firing of a gun had once been an arranged call between them. At three hundred yards' distance Johnson watched the figure of the other man. Was the shot a signal? A signal at that small distance? No, it was not a signal. Johnson's alert face relaxed in to the stupor of understanding the meaning of the shot. The gun had become an enemy. He stood still daring Phillips to fire again; derisive of the miss. Thirst-maddened, he had thought before that Phillips had wanted the gun to kill him; Johnson waited longer to see what would happen after Phillips disappeared behind the rock. The air and the trees were alive with the scrambling, chuckling nervous noises of the disturbed birds in their flocks. He glanced at them as he waited for Phillips to appear again. The trees were electric with wings.

He had left Phillips the gun. They could not say he had left the man without the means of defence; rather, he (Johnson) had stripped himself of defence. When he was with men his guilt was there—the illusion of seeing Charles Wright in the cave—but, alone, his luck came back. On his own, suffering cut deeper with self-infliction, and reward was greater. On his own, he would find water. On his own, his luck would return like merriness to his eyes, and he could do anything. Let Phillips come out from the rock and fire again and see that Johnson was impervious when alone.

But as Phillips did not come out again, a sullen depression and sorrow overcame Johnson. He felt the enmity of Phillips with wonder and, in his tortured brain, there were the vestiges of their common pity for each other. Johnson slowly turned his back upon the unlucky rock and the shamed man, instinctively not wishing to see his shame, and went forward to the water. Water, if there was any, might be on this side or on the other, but he went forward, forward and forward and the knife in his weak hand hacked with listless habit at the bushes and the trees to mark the way back.

When one imagines the hour before death it seems to be a time when the mind will ask, 'Why have I come here to *this* place to die? If I had known yesterday I would have eluded death by taking a journey to another place. I would have altered the events of my life, taking different turnings this time from the ones I took, and so I would have changed everything. A different course would have a different end.' One imagines dreaming the ancient platitudes: 'Why *now*?' and 'Can't I go back to settle this matter?' and so running a hand over the events of one's life with a pity for the creature who led it. But there may not be time or calm or consciousness. There may be no sad, luxurious survey and no meditative wooing of the end.

Phillips went back to the shadow of the rock to die. But he raved. Reminiscence became chaotic delirium. He muttered and shouted as his hands clawed at the rock. No orderly stream of pleading memory came back, but his mind rushed back to childhood as if he had been caught up in a burning ball and hurled there. His spirit shrivelled to the dimensions of a child's small frenzy of life and his clawing hands were upon his mother's breast, his lips parted to drink.

While he was lying there and Johnson got further and further away, the noise of the birds and the insects became furious and all the heavy-lobed leaves of the trees were loud with them. The macaws shouted and then the clockwork bawlings of parrot, toucan and cockatoo. In sudden silences these birds would take flight in numbers from one tree to another and start their din in a new place, while far beyond the place where Phillips was lying dense flocks went up and whirled around in indolent kite tails a quarter of a mile long. The insects matched the birds with shriller, more metallic sounds, bell-ringings and twangings, and the whole wild orchestra seemed to be working itself up to a climax which was not reached even when the sun went down.

Then, before sundown, nearly all the sound ceased and forest and scrub were held in an overwhelming silence. Here and there some great grasshopper hacked away, but he was an isolated worker, a mere stone in the depths of silence. The stillness stared and if a small bird now got up in flight it seemed to send a shadow over the breadth of the country. The warm air was thick and without movement except when, now and then, a noiseless and heavy wave seemed to pass through it like the breathing of a man, asleep, and when it had passed it did not relieve the burden of the air but rather added to it.

On this evening the sun was hidden before it went down. A line of dark cloud had appeared as suddenly as black heads of marching men over the horizon and then came in sultry battalions to meet the sun. Its last rays struck out for one last, lucid minute on the edge of the mass, and when they went many great trees in the scrub stood out, hard and particular, against the cloud. Presently the farther fields of the silence were frayed away by a murmur like the sound of the sea, the murmur grew into a far-off seething which came forward in invisible leaps and breeze shot out ahead of it, turning up the startling undersides of the leaves. And on the heels of the small breezes came the force of the wind, full, warm and powerful.

Restraint broke. A howl came out of the air and the trees bent over. They seemed to run before the wind, bowed double and streaming. The leaves flew flat and the branches ran out under them, and in the half-darkness of the cloud the leaves were no longer green, but grey. For tens of miles the forest leaned to the storm, the tall palms blowing like feathers in a hat and the whole scene like the flight of a mob.

Lightning had been twitching on the horizon but now the first fork split nearly overhead and the first peal went off. The flash and peal again, like beaten pans, the bush flaring, until the rain fell. It throbbed in blood-warm drops at first and then was hung suddenly in the sky like a solid curtain, billowing, blowing, and the dust and steam from the hot oil smoked at once where the rain swept, so that soon the clear sight of everything was wiped out.

Phillips had crawled to the doorway of a house of many pillars and balconies and there, a little man only a few inches high, in top hat and without coat or umbrella, he sank on the short flight of steps and clamoured to be taken in. This was his dream when the storm awoke his drenched body. He thought he must have died and that all the fables of hell had become true. He got to his knees. With a cry he opened his burning throat to the rain, he crawled to a corner of the rock which had become a gutter where the water streamed and he drank. He was down on his knees, his face pressed against the rock to catch the sand-reddened water. And from the rock he turned this way and that in a mad attempt not to lose any drop that fell around him. He coveted the water that dripped out of his hair on to his cheeks, that drenched his clothes and ran down his back. He gasped and talked aloud, calling out names to the water, and he laughed at it. Exhausted as he was by his frantic efforts to drink he could feel the first shivers of renewal to his life.

And as a little of his life came back he dragged the pans of his equipment out and with trembling hands set them to catch more of the water. They flashed in a way that terrified him when the lightning lit them with its methylated flame, but he groped among them panting and saying, 'Stay there! That's it! Fill up!' He laughed at them as though they were dogs or children he was training. He himself was crawling on his hands and knees like a dog.

Slowly, as life revived, he remembered his situation. He pulled his equipment and then the gun towards him and fell back dizzily against the rock in some inches of shelter. He began to sing and shout with joy. He crawled out and drank some water out of one of the pans and sank back again. Old fears came with the new life and he huddled in terror of the storm; and, when he had mastered or endured this, there was fear of the night which had come. And he feared for Harry. For how many hours had gone by, or even, how many days?

He searched for their torch and lit it. The rain did not slacken; it was a shining, hissing wall in the torchlight and lightning. He was soaked and swamped by it. The gutter stream had become a foaming waterfall. There was no shelter where he was. He was crouching in a river.

'There is always something to do. Before you need to fear there is always something to do.' If Harry was not there, there were Harry's words.

'The cave!' he cried.

The cave. How to get to the cave, save the equipment and get to the cave. What strength and mind he had became concentrated on these objects. His pride rose with the tempest. And then the long struggle began. The rain beat him back to the rock, mud and lumps of earth were breaking from the higher ledges now and the whole world—the small world in the small circle of his light—had become a vertical flow, a wall with no hand-grip. He searched with the light for a small ledge of shelter from the cascade and saw such an eave a few yards from him.

One was driven back flattened, breathless, between a wall of rain and a wall of rock, but he fought for it. The pans he left, but he dragged the equipment with him to the ledge. Again the flash of light upon a scene which had dissolved into spouts and waterfalls. They broke in beards and plumes from ledge after ledge in the light of the torch and down came the gravel and stone with them, flying out as if a man were hurling them from the top of the rock. He took the light away from a phantasmagoria so unnerving and kept it to the next step, for it seemed as impossible to live among those spouts ahead as under the fall of a Niagara.

Time passed in this scheming, fearing and reckoning but he had no knowledge of time. He was shifting his load, groping his way, sheltering where he could.

In the darkness he did not know that he was going the long way round to the cave and that, had he taken the opposite direction round the rock, he would have found it only a few yards away.

The peals rolled their war away to the horizon, leaving the rain windless. Lost in the rock face, bruised and drenched, he thought that the cave must have been another of Johnson's delusions. But the lightning in its retreat signalled the cave at last. It was high in the rock, a large hole half the height of a man with water spouting out on either side of it. He climbed up digging his hands into the mud on the boulders.

It was long before he discovered the size of the place. The rock roof was low inside and the depth at first seemed little, the air was warmer than it had been outside. He sat down and the water poured from him. He was too exhausted to explore. While the struggle was on he had had strength and nerve for anything, but soon he found that the awakening to life was an awakening to misery. The rain made the intolerable whine of machines on a factory floor.

The expedition had taught Phillips many things. It had taught him always to be active. The torchlight was strong though it had only one or two hours' life in it, and, careful not to examine the cave yet and so start fears rising, he concentrated on the little circle near him, and unstrapped the equipment he found his sodden matches and laid them out on the dry floor. He rang out his clothes and then put them on again. His mouth and tongue and lips were aching and he could not easily swallow, but he took some of the sodden meal which had become like wet dough in the sack and tried to eat it. It made him sick. Having resigned oneself to a torturing death once, one is not going to give it another chance without a struggle.

He started to talk to himself as he crawled about his muddle in the floor of the cave. He was with quiet feverishness making a world for himself which had no past and no future. The thoughts: 'I shall get pneumonia because I cannot dry my clothes and shall have to wait until sunrise to dry them,' or 'What is Harry doing in this? Supposing he doesn't come back tonight?' were pushed away by a search for the rest of the food, for a brandy–flask which did not exist and the hundred little gropings among his muddle. And all the time he was thinking, 'Fire. How can I get fire? I was burned nearly to death a few hours ago and now I can't get a light anywhere.' He began to murmur to himself, 'Give us a light. Give us a light.'

'Give us a light, someone,' he said aloud.

He flashed the light round the cave.

'Give us a light!' he shouted.

His voice filled the cave and there was a small and remote echo after it. He knew it was an echo but he called:

'Who's there? Come out and give us a light.'

The far corner of the cave was not closed and, when he crouched to look, he saw there was a rift which led to another compartment in the rock, very lofty at the opening. The flash of his light brought the bats down from the roof. It all so vividly recalled going to the pasteboard river caves of the Crystal Palace when he was a boy and all the 'horrid caverns' of the mountains of romantic literature, moments of tense reading of Scott and Shelley by the kitchen fire, that his fears had almost a comforting familiarity.

'Bloody homely,' he said. And 'homely,' chattered the echoes.

Weakly he started to sing hymns and like some famished gospel meeting the echoes chanted. He came out from the cavern to the hiss of the rain in the outer hole with a gratitude for the dreary company he had. They never failed his voice.

He was determined upon fire, to dry the things, eat and break the misery and fear of the night. Johnson and he had been on little more than the level of animals and, crouching in his hole, he marvelled at the persistence and tenacity of the animal under the degradation and flimsy exaltations of the mind—so treacherous and incomplete an instrument. How pitiful the naked helplessness of the animal and yet how ancient its cunning! The very pain that went through the nerves, and the fever that hazed the eyes and the blood, gave an intoxication to the instinct to live. He swayed through his acts with the sixth sense of an inebriate.

The storm abated and a second storm followed as violent as the former one but shorter. Those detonations broke and crumbled like organ music in the depth of the cave. And the rain went on. In the early lulls he crawled to the cave mouth and flashed his light and called, but there was no answering call and the only sound was the rain in the trees. It did not cease and all night the downpour went on, like the sea breaking upon his rock. He had the sensation of being a shipwrecked man on a raft tossing in blackness. No human response came to the call and no animal sound.

The torch failed. He groped towards the heap of dry sticks he had been collecting in the cave and knelt over it feverishly beating flints to make an igniting spark. Beating, beating, eyes stretched with hope of the spark, the sound like castanets in the cave.

'Oh God, give me fire,' he called out, but even in his desperation a little theatrical in the use of the word

The power of his own voice among the rock was too great.

'His bloody father!' he said more quietly to trick the echoes. And slowly from this he drifted into prayer which was merely a return to the habits of his childhood:

'Our Father which art in Heaven, Hallowed by Thy name.'
(Damn fool letting Harry have all the brandy.)
'Thy Kingdom come.'
Get it through quickly unless a fear or pain stops you and then
you have to begin again—
'Our Father which art in Heaven . . .'
It degenerated into anything that came into his head and was
never completed; bits out of Dickens from Harry's *Pickwick*,
imaginary speeches to Charles Wright and Harry, long detailed
explanations justifying his conduct, any word that would annul
fear. And he returned to hymns because his childhood was in
them. The flints were still in his hands when he slipped back
exhausted against the cave wall into agonised sleep.

The sun woke him. He had his clothes and the food out at
once. The steam rose from them as it was rising from the trees.
The earth was darker in colour, swamped, swollen and exhausted, and the low clouds were like dirty trodden snow, yellowish
with thaw and the sun lifting them. For miles the vapours were
rising derelict from the trees and the water poured, though in
lesser falls, from the rock. Darker clouds were blobbed over the
distances and closing out the sky. The air was heavy and
unbreathable with the steaming heat. It was extraordinary after
all their cloudless and dazzling mornings to see the cloud stupor
of the sky and the earth under it like some drunk, gross harlot
who has slept where she has fallen. The smell of the forest was
like the smell of spirits gone sour on the breath.

'The fool!' he suddenly said, for the way to make a fire had
suddenly come to him. He opened a cartridge, poured the
powder into a little heap by the sticks in the cave and he had
soon struck a spark large enough to start flame in the powder.
He was so busy with his triumph with the fire that he did not
notice that the sun had gone and that the rains had begun again.
They came with no warning, as if the sky had been tipped up
and the water emptied from it in a solid stream. He got the
things in and set to work to dry them again. The smoke filled
the cave and he was driven into the inner chamber.

In the past night Gilbert Phillips had touched the depths of human wretchedness and he had risen only a little way out of them on this morning. He worked painfully, listlessly, like a roped man and his eyes were as nervous as an animal's glancing with apprehension and hope at the land from the mouth of his cave. He looked up at every sip of the hot drink he had made and at every puff of the distasteful pipe he smoked. His mind was divided between Harry and England. At the next glance Harry might be there and with *every* glance Phillips played the old trick of a man alone: the trick of populating the country around him. Houses appeared among the trees, rows of suburban villas, traffic and people. He made ideal colonies. He dreamed for himself a tribe. All these things protected him.

To act was vital but wearying. The sight of Harry's gun stirred in him some almost superstitious sense of duty and he worked to clean and dry it. There was virtue in the weapon. He was astonished to find that one cartridge had been fired, because he had no recollection of firing it. The memory of the impulse to kill was distorted in his mind into a feeling of power in possession of the gun and another feeling of equality with Harry. This was very natural, for he had realised one of those curious, half-conscious desires of the march, one of those silent covetings of the traveller, in getting the gun out of Harry's hands. But the fact remained that it was Harry Johnson's gun and this knowledge expressed itself in a way that comes instinctively to man when he is living on the last line of primitive survival. On wiping and cleaning the gun there was a remote memory of a belief that goodness came out of it into himself and that the benefit of the act would reach Johnson too, wherever he was.

If Johnson had come back his practical, experienced eye would have seen that Phillips was doing most things badly. A belief in the possession of Harry's golden luck was more urgent to this romantic than a careful counting and drying of stores. Johnson was one of those men who are not conscious of the forces that impel them but suppose merely some practical end. The impatient romanticism of Phillips was incapable of much careful attention to concrete fact; though he had helped plot the maps and had often studied the problem of direction with Johnson, he had really no very clear idea of where they were and how they would get out and he did not understand in the least that what Harry called 'luck' was really the result of careful thought, planning and common sense.

The gun was, for Phillips grovelling in his cave, the symbol of this luck of Harry's, his leadership and courage. The hope Phillips had was nevertheless small. Anxiety about Johnson's fate could not be discussed. The rain though strong was fitful, and when it stopped he already expected Johnson to appear like a man who has been sheltering in a doorway near by. When Johnson did not come, the dread increased. He had pushed out of mind the prospect of another night alone, but now that prospect became more likely. He picked up the talismanic gun, gripping the barrel hard, and said:

'I couldn't stand another night. I couldn't stand it. I would sooner shoot myself with this gun.'

Still, there was time to hope. Only half of the day had gone, and now once more the rain had stopped, and the hot blue body of sky thrust limbs and shoulders through the cloud. Voluptuous and carnal in the sun was this Rubens-like sky, to the man of shivering, pain-thronged body in the cave. He sat outside on the rock to feel this warmth in his blood, though the insects teemed like another rain upon him.

He dozed and woke and dozed again in the sun. It was in one of these jolted wakings that he saw a tall spindle of smoke rising from the bush. The distance was hard to judge. It looked a mile away. The smoke was thick and the colour of a vein in the hand, easily seen against the dark vegetation. The fire was in the part of the country through which Johnson might have passed.

Phillips climbed higher on the rock and saw no other fires.

It was a new fire, piled high with damp stuff, a signal no doubt. Harry had no gun to signal with so he had lit a fire! He had lit a fire on high ground to show he was on his way! He was held up there, perhaps by a river in flood; perhaps he, too, was water-logged and ill in some cave. Perhaps he was lost and was asking for guidance.

Phillips hurried to build a fire of his own on the rock with the dry wood he had saved, and piled up more wet wood from the bush. He made a great fire and the smoke went up thick and white and bold. Still the unchanged smoke of the other fire went up straight against the trees and cloud.

But a change in the light showed the nearness of the fire to be illusory. The smoke was more than a mile away, it might be two or three miles and there was no knowing what lay in the middle distance. Phillips went up again to the top of the rock and this time he saw that the hill lay under the sun, the direction Harry had taken not much later in the afternoon. But now there were two columns of smoke.

Johnson would have lit only one.

At once Phillips climbed breathlessly, heart-bursting, from the rock to his own fire and began breaking it up, throwing the burning sticks around him. He kicked at the embers and wished to pull down the smoke which, in agony, he saw escape him into the sky. Those columns were not Harry's signals but what Harry had always hoped for and he, night after night, had dreaded in the march. The rains had come and the Indians were returning to the country they had deserted. Gilbert picked up the gun and stood back against the rock looking at a bush in which every shadow seemed the shadow of a crouching man and every dapple of sunlight a face brilliant with deceit.

CHAPTER SEVENTEEN

HE had to make up his mind and he was a man who could not make up his mind. He either sank back paralysed or went forward to his act on the uppermost instinct of the moment, dragging his quarrelling mind after him. He would go to heaven in dispute with hell and to hell in dispute with heaven, like old Captain Mommbrekke, Lucy's father, in contention with his conscience. He would sing songs in church and hymns in a brothel. He would go on a journey he did not believe in with an objective that was absurd. He would die when he ought to live and, damnation take it, he would—if he survived—live when he ought to have been long ago dead. And now, if he stayed still and waited by the ruins of the fire he ought not to have built, he knew that it would have been better to go on,—and if he went on towards that distant smoke, he would have been wiser to have stayed.

For—it was clear to him—the Indians were on Johnson's route. They must have seen him. Or he must have seen them. Johnson had had a desire for the fires of the Indians as he might have had a desire for death. He would have gone to an Indian fire sooner than to water. The chances that they had missed each other, Indians and Johnson, were infinitesimal. Two men might miss each other, being on the opposite sides of a wood, but not if one of them was an Indian.

Well, he (Phillips) had populated the trees, his instinct had dreamed a tribe of his own and here was his tribe.

He waited for an hour to see what emissary would come from those fires, for his own must have been seen. It is the curious condition of profound fear to end by desiring the thing feared and Phillips, fear-nailed to his rock, was drawn out by his own magnet. His hands trembled. At last he was indignant that no one came, angry with Johnson for not coming if he were by some chance *not* in that camp, angry with the Indians for

212

remaining there. Once or twice he considered firing a shot but he thought:

No, now is not the time. A little later, when the time comes, I will fire.

And when will the time come?

I do not know. But there will come a time when I will fire then.

The anger lit by his anxiety grew. He paced up and down the rock. He went to the cave and looked at his things. Useless, spoiled, not a day's life in them. (This was quite untrue, but was how he thought about their stock.) He looked at the stuff in the manner of one who has finished with it. This is the end of one stage; let another begin. He understood the nausea he had before every minute of the past, the disgust with his kennel-cave. I am not going to die in a hole or live in a hole. Why the bloody hell don't they come? Why doesn't somebody come? Die here where I suffered? Endure another night in this place? No, any place but this place. Some other place, I beg.

They did not come.

Now he wanted to stand on the top of the rock like Samson taunting the Philistines. He wanted to shout to them. He did not shout but merely muttered and as the anger swelled him, and he stood gun in hand like a fuming sentry on his watch, he felt the luck of Harry in his hand and the courage of Harry too. If Harry was in their hands he must go to him, he must rescue him, surrender with him, parley with him. If Harry was leading them now in peace towards him, he must go and meet them. Whatever it was there must be an identity of fate between them. The love for Harry which had grown out of the love of Lucy demanded this.

'Take care of him for me.'

And what hatred and jealousy of Harry there had been demanded it too, if any were left after the forgotten mock-murder of the day before:

He must not die away from me and let me die, but if there is to be dying we must die in the same place.

The mystery of the father is made clear by the virtue of the son.

And I do not want to be alone.

He had taken his bearings now; with his anger and his joy he relit the fire for a mark, strapped on his pack and climbed down the rock into the swamped bush. He was going to Harry. To Harry and Harry and Harry! 'Two men are better than one.'

It was what he had always wanted: to be the conqueror and rescuer; and though at once he saw the bush was sodden and great eyes of water opened in it, waist-deep streams had to be crossed and on all sides the trees were awash and water-birds had come down to these lakes, he was not intimidated or tricked by sudden depths. Lucy was with him talking even as he hauled himself up the mud of a bank or stopped, on rising ground, to regain the sight of the guiding smoke or guess with cunning his direction. They were going to Harry. She was to near to him, sometimes beside him and sometimes in himself, so that he talked and laughed with her voice. There could not be a closer possession. There could not be a greater unity of the soul and from their unity this lucky courage came.

The sighting of the smoke had become more difficult and there were long stages in which it could not be seen, or when it had become far more distant. He though he was there on the next rise but there was a vale and a rise beyond it. A shallow valley presently opened, descending across his path, a valley which widened and deepened on the south-eastern side, away from his route, until the forest thickened on the rising hills beside it. A great panorama of forest lay beyond this rift far below the plateau.

He stopped in amazement at this sight. He was looking down upon an endless sun-shot sea to trees without break or landmark on it. He and Harry had been mice nibbling at a continent. But he stopped too, because it was clear that a man in search of water would inevitably have descended this valley where now the swamps drained into half a dozen streams.

Phillips crossed the valley with his first doubts and they broke the unity of his spirit. The company of Lucy left him and now he floundered. He began to walk warily round the flooded stretches

instead of splashing through them. In the tree belts he sank into a
rotten floor of fallen trunks piled waist deep; sometimes his foot
would go through the infested shells of the tree-trunks and he
was up to his ribs in the wreckage of the forest floor and the
biting flies went up in black clouds around him. He was going
uphill now, panting in the green stench.

In the struggle it seemed to him he became huge and more
vulnerable at every step, grotesquely naked to watching eyes.
Where were the eyes? Where were the listening ears? Did they
hear this crash of branch, this curse as he slipped? From one log
he fell full length into the bark mud; he half smiled, apologising
to invisible eyes for unpresentable appearance. Could it be
explained that Englishman, no ulterior motive, looking for
another Englishman, had fallen down, got muddy, meant no
harm, would put all right in a minute? Could it be explained
that he was not alone; oh no, advance-guard of whole nation,
you dare touch us! You let that man free! Please understand.
Civilised men in love with same woman lost in your country,
searching that gentleman's father. Appeal common human
feelings, blow your brains out if you move. He was almost at
the top. Ten yards before the hole of sky ahead became level
ground.

'Christ!' said Phillips.

He clutched at anything and then, panting, exhausted he was
there. Headlong there, for down he went again in the opening
flat on his face and the gun went off.

'Did not intend that, but since signal has been made . . .'

They were waiting for him there, fifty yards away—the
Indians. They looked at him and ran, vanished like deer into the
bush.

They found him on his hands and knees. The Indians came
first and then the three Germans and the Dane. They asked him
who he was and, when he answered, one of them replied in slow
bad English.

He said he was looking for his friend. Two of them laughed
loudly at this. They were short, thick men. Two were bald but
the Dane had thick fair hair.

'Don't laugh at him,' the Dane said.

'What friend?' they said.

Where did he come from?

They saw him stand up and then his legs went under him.

They carried him into the camp. Twenty Indian porters were with the Germans and they had mules. The Indians had built thatched shelters on poles. The camp had suffered from the rains.

They gave him brandy and waited.

'We saw your fire,' they said. 'Our men were frightened. They would have killed you but for the gun.'

'It was lucky having the gun,' he said. 'It belongs to my friend.'

'Don't disturb him,' the Dane said.

They left him alone. He was lying in a hammock under the thatch.

EPILOGUE

CHAPTER EIGHTEEN

I READ the article,' she said. 'You don't think it is he? You don't think he will ever be found? It just won't ever be known?'

'If you had seen that country!' he said. 'Cut it down, every plant and bush in it, and the whole place would be covered again in six weeks. It's like the sea. It drowns you.'

'Do you think he was drowned?'

'I think he was. I certainly don't think he's alive. The Germans who searched for him thought he was drowned. He must have gone by that valley, the one I crossed, the one I told you about. He was looking for water. Well, he must have gone down that valley along the river-bed. All that country was draining into it. And he had no gun, you must remember that. He left the gun with me.'

The gun was standing in the corner of the room. The fire in the dim January afternoon made a seam of light along the barrel. They had talked themselves to silence. The street air hummed against the windows and there was the gritty sound of London rain. He was still not used to the sounds.

'Well, Lucy,' he said, in another voice to change the subject, 'when is it going to be?'

She was sitting beside the fire.

'What? Oh, you mean the baby?' she said. 'In April.'

'And you're happy?' he said, tipping back his chair.

'Yes, very happy,' she said absently.

'I'm glad you're happy,' he said.

'Gilbert,' she began again for she would not let him change the subject. 'It was a mistake. I knew it was a mistake. I knew this would happen.'

'It was not your fault,' he said, misunderstanding her.

'I don't mean it was a mistake that we loved each other.'

'Harry and you . . .?' he asked, jolting down the legs of the tilted chair. For she might mean himself. Even now she still might mean this.

'Yes, of course,' she said with a hard, cool dart of her voice. 'I mean it would have been a mistake for us to have married. He would have killed any woman. That doesn't mean he wasn't nice.'

'He would have hated her for being a woman,' Gilbert said.

'He wanted women to be men,' she said. 'He really wanted me to be a man. In his head he did.'

'He was in love with his father,' Gilbert said.

He looked into the fire and the coal was like a rock and the glow below it like the glade in the fierce sun. A man was walking away down the glade without pack or arms. Phillips remembered this. He remembered only the back and head of the man in the sun and nothing more. The sorrow of that last sight had gone but an unaccountable feeling of shame and triumph returned—shame that all attempts to find Harry Johnson had failed, the shame of the day when the Germans gave up and broke camp, saying, 'He is dead. He must be dead. We can't stay here and we have looked everywhere. We have been to the rock and the river. It has risen fifteen feet and is mile wide now and still rising. He cannot possibly have lived. He was right about the river and the water but if he got there that night he could not have got back.' Phillips looked at the trees. Where in those trees was Harry lying? Or standing? 'I cannot leave him in those trees. Shall they leave me behind to look for him? They can give me stores and an Indian and I will stay here.' It was a dilemma. The Germans were going down the valley into that country which had looked in the sun like a Promised Land, down the valley Harry must have taken, and they were going beyond where he could be because they had boats. He could go with them but in a day they would pass Harry's utmost limit and then, following their river, would ultimately reach the coast. Or he could stay there, daring and seeking the identity of fate, waiting with an Indian, and then, if Harry did not appear, he and the Indian would return to the branch of the main river by which the Germans had come ten miles away.

The Germans looked at him with pity and talked among

themselves in their language. They knew about him. They knew about him not only by what he had told them but from Calcott. They had passed through that town. They expressed no opinion but there was the hint of a smile on their lips when they talked of the importance of using modern methods. How many weeks was he on the river? So! They had done it in half the time. They talked of the fast motor launch they had brought down from New York. They had heard a man had died. Now two men had died.

There was a luxury in the pity of the Germans, something to accept with one hand and to reject with one's secret pride. His jokes were translated. The Germans laughed. He was a witty fellow.

But the rains were upon them. Every day the sky emptied and the country steamed. One of the Germans fell sick. That—if the decision was not taken long ago in this cradle—decided Phillips.

'I'll go with these people. They have been good.' They seemed—or did he image this?—to despise him for his decision.

They set off, a straggling procession, the Indians with the boats on their heads, heads bobbing like corks through the tall grass. For a quarter of a mile the line was drawn out, here three or four men, there one alone and then the group walking by the litter on which the sick man was carried. The cursing when, in the bottom of the valley, they came to the headwaters of the river! The natives who went back!

'You know your friend, I mean your other friend,' said the bald, dark-eyed German who spoke English. He tapped his feet to explain. 'Boots,' he said. 'It has become a political affair because a Brazilian was thrown into gaol on a false charge of stealing his boots. They said it was a manœuvre of the clerical party.'

Two promontories of the forest rose above the valley and he and the Germans were walking at the foot of them in the steaming greenhouse shade of the great ferns. He smiled, and their talk had covered Phillips' retreat and the unreasonable shame of retreat.

'We couldn't do more than we did, Lucy,' Gilbert said,

looking up from the fire. 'I could have stayed but it was futile. I really think it would have been futile.'

'It was brave of you to go on too,' she said. 'You had no choice.'

'I hadn't. I can't believe it yet. But I hadn't,' he said.

'And in the end,' he said, wishing to God that every word he spoke about the expedition was not haunted by the forced note of self-justification, 'we passed by Charles's country. It was no different from anywhere else. It might have been anywhere. Just trees. Nothing but trees and trees. I was glad about that. It was like finishing Charles's job. It wiped out something. I told you?'

She nodded. The bones of her face were more prominent, the eyes deeper set. She looked as though she had gone back to her religious ways which gave her a heavy expression of obstinacy, slight illness and distance. She was only pretending to listen to him.

She was thinking all the time of the shock of seeing him. She did not wish to go over the past, she did not wish to dig up the dead. She had known for a long time that he was in England and she had avoided him, put him off in her letters. Couldn't he see that everything was different for her now?

But when he said, 'That wiped out something,' her heart said:

'And my child, doesn't that wipe out something? You cannot tell me *now* that I ought not have married. I had to marry. I had to marry when he went. I was thinking as the steamer sailed and Harry was waving from the deck, how slowly the ship moves from the dock! Make it quicker so that we can be out of each other's sight at once because I must marry before I get the chance to catch the next boat and go after him. I must chain myself down. I must marry someone quickly. And when I had married I thought, "This hasn't chained me." Marrying isn't enough to wipe out that man, I must have a child.'

She was defending herself. She was defending herself from her grief. The wound Harry had made in her was intolerable. They had loved like robbers.

One had to annul that feeling. One must smother the old with the new. If one can spin enough words they will fall like a web over the old feeling and change it. She had a child now. That put Harry and Charles and Gilbert in the past. The child, with new demands, wiped out the betrayal of marrying. She saw Gilbert looking at her and she could feel her growing child, then her husband, then all the months of the past year, the exchanged letters placed between Gilbert and herself. He looked older. The fair hair was bleached, the tanned face was leaner in the jaw. But his voice was the same and his manner had not changed; too disturbingly they recalled the year.

'Why are you frowning, Lucy?'

'Was I?' She laughed. How observant he was. Lost in his strange world of his duty to Charles, his duty to Harry, the subtleties of honour and obligation—and yet so observant! Her parted lips closed with the stern, placid line they now so frequently chose.

She was frowning because when she looked at him, heard his voice which seemed to attack her with his story, she could not believe she had been his mistress. She had been fearful of coming to see him because she had felt that he would bring not himself only but the intolerable presence of Harry to her. She feared that she might discover she loved him because of this.

But she did not love him. He brought to her this ineluctable knowledge that where once was feeling there could be nothing.

'Were you surprised when you got my letter?' she asked. 'The one I sent to Rio? Were you in hospital then?'

'I've never been surprised at anything you did, you know,' he smiled.

'I hadn't heard then, not even about Charles,' she said. 'We'd been married a week when I heard. It was awful and frightening. I didn't write to Charles. Harry felt badly about him, you know, and as I wasn't writing to Harry I thought I had better not write him. It would have worried him.'

'You didn't write to me,' he said. 'Is that why?'

'I suppose so,' she said.

'Did you tell Harry you wouldn't write?'

'I didn't say anything. It was over. We agreed it was over. I thought I would wait till the expedition was finished, to give him time to forget about it, and then'—her grave voice became quicker and lighter with the gaiety he remembered—'just write him a very friendly letter.'

She valued the art she had of concealing feeling. With Harry she had dropped the pretence fatally. Never again. She gazed at Gilbert out of her mists and fictions. He was the same man, but dead to her, more dead even than Harry and Charles were.

But she was not the same woman. Or she was not the woman he had carried in his head with Harry. He could not even recall the times he had seen her with Harry except that last night of Harry's in England, when he had seen her on the stairs in her white nightdress calling him. The slave in her had gone, who followed Harry picking up his clothes, collecting his papers, making arrangements for him, agreeing to do all the things she disliked because Harry asked her—did not even ask but said, 'this is what we are going to do.' He could not recall those times. The new fashions changed her, the height of hat, the new dress. They gave a brisker, more actual, appearance—a woman who sits at a bare table having swept away the old papers.

They must get up and go before they had to admit they were bored with each other; and bored because each reminded the other of too many dead things. It was necessary to be bored.

He said, going to the corner of the room where the gun was and bringing it to the fire:

'It was Harry's,' he said. 'He left it with me. I told you. I should like to give it to you.' He was embarrassed. What an absurd gift to a woman!

He wanted to get it out of his sight. He did not want to be reminded of the day of the rock, the last day he had seen Harry.

But she thought, 'This is too much!' She had not told her husband anything.

She and Gilbert were trying to hand each other their guilt.

No, you must keep it,' she said. 'It ought to be yours, Gilbert,' she said.

They stood by the fire hesitating. She lowered her eyes and then, raising them, again said gently, half laughing:

'But I couldn't take it now. It's heavy. I couldn't go through the streets with a gun. Some other time ...'

'And I must go, Gilbert dear,' she said.

Suddenly he wanted to laugh at this ridiculous situation.

He went to the door and switched on the light. He saw then that at one time in the difficult talk she had wept. She held out her hand.

'Don't worry about the newspapers. Newspapers make me sick,' he said, picking up the paper from the table. It was she who had brought the paper; indeed the paragraph that she had read there had made her come to see him. He looked again at the paragraph. 'They'll find him twice a year after this. Just like his father. There will always be rumours,' he said.

'And he hated anyone even to speak of anything he did,' she said.

He went with her to the door. Nothing could have been more like the river and the jungle and the sudden squawk of birds to his unaccustomed sight and ear, than the street light-daubed in the rain, the impenetrable forest of lives of people in the houses and the weird hoot and flash of the cars. He went back to his room, wishing for a friend.